George Charles Frederick Husmann

Grape Culture and Wine-Making in California

George Charles Frederick Husmann

Grape Culture and Wine-Making in California

ISBN/EAN: 9783337327866

Printed in Europe, USA, Canada, Australia, Japan

Cover: Foto ©berggeist007 / pixelio.de

More available books at **www.hansebooks.com**

GRAPE CULTURE AND WINE-MAKING
IN CALIFORNIA

A PRACTICAL MANUAL FOR THE GRAPE-GROWER AND
WINE-MAKER

GEORGE HUSMANN
NAPA, CAL

SAN FRANCISCO
PAYOT, UPHAM & CO., PUBLISHERS
204 Sansome Street
1888

PREFACE.

A book, specially devoted to "Grape Culture and Wine Making in California," would seem to need no apology for its appearance, however much the author may do so for undertaking the task. California seems to him, at least, as "the chosen land of the Lord," *the* great *Vineland;* and the industry, now only in its first stages of development, destined to overshadow áll others. It has already assumed dimensions, within the short period of its existence, hardly forty years, that our European brethren can not believe it, and a smile of incredulity comes to their lips when we speak of vineyards of several thousand acres, with a product of millions of gallons per annum.

But, while fully cognizant of the importance of these large enterprises, it is not for their owners that this little volume is written specially. The millionaire who is able to plant and maintain a vineyard of several thousand acres, can and should provide the best and most scientific skill to manage his vineyard and his cellars; it will be the wisest and most economical course for him, he can afford to pay high salaries, and the most costly wineries, provided they are also practical, would be a good investment for him. We have thousands, perhaps the large majority of our wine growers, however, who are comparatively poor men, many of whom have to plant their vineyards, nay, even clear the land for them with their own hands, make their first wine in a wooden shanty with a rough lever press, and work their way up by slow degrees to that competence which they hope to gain by the sweat of their brow. Of these, many bring but a scanty knowledge to their task; and yet it is from these, who cultivate their small vine-

yards with unceasing interest, and are willing to watch their
wines with the grestest care that we must expect our choicest
products. To help and serve this army of patient toilers, in
whose ranks I have labored for forty years, here and in Missouri, with hand and brain, is the object and aim of this little
book; I can fully sympathize with them, because I had to
gather what little knowledge I may have, piecemeal and by
hard practical experience in an almost untrodden field, and I
wish to save *them* some of the dear bought experience which
I had to pass through.   If its pages become a practical guide
for them, by which they can plant and cultivate their vineyards, prune and train their vines, erect their wine cellars
when they need them and are able to build them, and make
good, drinkable and saleable wine, my chief object has been
accomplished.   To do this, I intend to be as concise and
clear as possible, use no high-flown language, and avoid scientific terms as much as possible; talk as the plain, practical
farmer to his co-laborers, and confine myself to simple facts,
gathered from my own daily practice as well as from the
practice and counsels of others who have labored long and
successfully in the same cause.   None of us are infallible,
and the best way to gain knowledge is by exchanging ideas
and experience among ourselves, comparing notes with each
other.

And this is especially necessary in each neighborhood,
each valley and its surrounding hillsides in this, the brightest
and most bountiful, but also the most diversified and variable
State in the Union; where the climatic conditions as well as
the soil change as quickly, according to each location as in a
kaleidoscope.   This makes it all the more necessary, that
the vintner select his climate and soil carefully; and again,
that he chooses such varieties as are adapted to his soil and
climate.   Then the climatic conditions will also materially
affect his operations in wine making, curing raisins, etc., in

short no man can hold fast to one invariable rule in this State, but must vary his operations with the location, the soil, the product of each season, and the climatic conditions prevailing during his operations, and which may vary every day during the vintage. And therefore this will be a *California* book, first and foremost. I am fully aware that French and German, Spanish, Italians and Portuguese, have many eminent men who have compiled the experiences of centuries, and from which we can learn a great deal, especially in making and handling wines. But while I value it highly, and am willing to profit by it, yet all the conditions are so different there, that they can be no safe guide for us. Our rainless summers, the character of our grapes, which always ripen, and are heavier in sugar, while they may lack in ferment and sprightliness, will necessitate different handling, and I believe that three seasons of active experience here, will enable a man with good sound judgment to make a more perfect wine from California grapes, than twenty years of practice in France or Germany. He has nothing to unlearn, is free from prejudices and antiquated methods, and is therefore more likely to succeed, than the one who comes to the task with the preconceived notion that he knows everything, while in reality he has to take lessons every day. Therefore, though I will gladly use some foreign experience, and give due credit for it, this little volume will not be a foreign compilation, but a practical record of California experience, in the vineyard and wine cellar. That this may be concise, useful, and offer such assistance as practical men may need is my highest ambition. The reader must not expect infallible doctrines, nor impractical theories, but plain rules, variable according to circumstances, and given in plain language, without poetic fiction or privilege. I shall try to deal with, and confine myself to plain facts of every day occurrence.

But while its principal aim shall be to become a guide to

the beginner, I also hope to make it interesting enough for
those who count their vineyards by the hundreds and even
thousands of acres. I hope to give a true and full picture of
this giant industry, none the less gigantic because yet in its
infancy, and which owes so much to their enthusiasm and enterprise. They will furnish the wines for the million, and it
is truly a noble and proud task to furnish to every laborer a
sound, cheap and palatable wine, at less cost than tea or coffee. We want their assistance to make this great nation temperate, convert them into wine drinkers, instead of drinking
so called brandy and whisky, the banes of so many otherwise
happy households. And for this object we may safely claim
the help of the ladies also. I am proud of the active part so
many of them have already taken in the viticulture of California. There are not a few of them who successfully manage
vineyards and wine cellars of hundreds of acres, and hundreds
of thousands of gallons. May their numbers increase, and
they become our helpmates in this as in every good work. I
am sure that I commit no indiscretion if I mention the names
of Mrs. Kate F. Warfield and Mrs. Hood of Sonoma Valley,
and Mrs. Weinberger of St. Helena, as among the foremost
in this State, while the illustrious example of Madame la
Duchess de Fitz James, who has already replanted about two
thousand acres destroyed by the phylloxera, on her estates in
Herault, France, has given the results of her experience to the
public in several books, and inspired new confidence in the
industry into the poor despairing peasantry of that district,
who saw their only means of livelihood failing them, should
not be forgotten here. I use the term "illustrious" advisedly,
not in reference to her rank ; for I am Republican enough to
have little regard for the accidental privilege of noble birth;
but a woman who does such noble work, winds a more shining and lasting crescent around her brow than monarchs can
confer, and birth bestow, and is worthy to be counted among

the benefactors of mankind; I am proud of the privilege of being her correspondent, and thus acknowledging her noble efforts in our cause.

This work was commenced in June, when the greatest rush of vineyard work was over, written in my cabin in Chiles Valley, surrounded by vines, where practical reference could be had every day and hour to the operations necessary among them. As it progressed, I became convinced more and more of the magnitude of the subject, and the impossibility of doing it full justice in the space of a few months, and a few hundred pages. I hoped to complete it before the vintage, so that it could be of some use perhaps during its progress. But unavoidable delays have drawn it out to the end of the vintage, of this truly abnormal year, abnormal in its late and destructive frosts, its hot winds during summer, causing a great deal of coulure and its unusually hot weather during the vintage. It has been one of the most difficult seasons to handle a vineyard and wine cellar, which will ever occur here, I trust, and has taught us many and severe lessons, among others the importance of thoughtful pruning, close attention to the growing crop, and diversity of varieties in time of ripening, so that we are not compelled to crowd the operations of months into a few weeks ; also the necessity of commencing the vintage as soon as the grapes are fairly ripe ; and of unceasing work during wine making. Vintage work commenced in Napa Valley about the middle of September, when it ought to have commenced a week sooner, and the hot and dry weather prevailing during its entire period even until now, ripened and dried up the grapes to a certain extent and thus fermentation became very difficult, especially in large establishments. I have availed myself of these practical lessons, I hope to the advantage of my readers, so that the delay may not be a loss altogether. Still, I am aware that it is impossible to do the subject full justice, and am far from claiming that I could have done so. Now, when my task draws to its close, I ask their kind indulgence; to them I must leave it whether it has been done well or ill. I can only say that I have tried my best to be useful to them and to the industry at large.

I cannot close these remarks without grateful acknowledgments to those who have aided so materially, by their workings and contributions, and without whose help my task would have been infinitely more difficult. I have drawn freely from the bulletins and report of viticultural work of Professor Hilgard of our State University, from the reports of our State Board of Viticulture, the valuable work of Mr. E. H. Rixford "The Wine Press and the Cellar," from the ampelographic dictionary of Prof. Hermann Goethe, and the writings of Mr. Chas. A. Wetmore. I am also indebted for courtesies and valuable information to Mr. J. H. Wheeler, our present chief viticultural officer, to Mr. C. J. Wetmore, the Secretary of the State Board, Mr. W. B. West, of Stockton, Cal., Mr. Horatio P. Livermore, Mr. H. W. Crabb, Mr. Charles Krug, Mr. J. W. Hale, Superintendent Barton Vineyard at Fresno, E. M. Maslin, Secretary State Board of Equalization; Mr. M. Denicke, Fresno; Mr. Juan Gallegos, Mission San Jose; Capt. J. W. McIntyre, Vina; Mr. Shackleford, Vina; Mr. Smith, Vina; Mr. D. M. Cashin, Secretary of California Winery and Security Co., Mr. Julius Dresel, of Sonoma, and many others. I only regret that I could not elicit a single satisfactory answer responding to requests for information from Los Angeles, as I was anxious to have the whole State represented. But the reply from all was, that they had no time to give the necessary information. So, if my information from there should prove meager and incorrect, I must lay the blame at their door, as it was certainly my desire to give full and true information.

To the press of the State in general, and especially our

local papers, our industry owes a great deal, and I have freely drawn from the information they give. They have always taken a lively interest in viticulture, and published all the information they could gain, giving due prominence to our calling, and I take this opportunity to tender them our grateful thanks.

Hoping that they and my readers will receive this volume with their usual indulgence and kindness, I remain

Their fellow laborer,

George Husmann.

OAK GLEN VINEYARDS,
CHILES VALLEY, NAPA Co., CAL.,
October 20, 1887.

CONTENTS.

Part I.—GRAPE CULTURE.

CONTENTS.

Part II.—WINE MAKING.

PART I.

GRAPE CULTURE IN CALIFORNIA.

CHAPTER I.

It cannot be expected, in a book which pretends to be no more than a manual for the grape grower and wine maker, that I should give a history of the industry in California. This, although no doubt it would be a pleasing task to note down its earliest beginnings and do honor to its pioneers, requires an abler pen than mine, one imbued with all the poetry of the subject, and with all the leisure to trace up their records, than can be brought into a practical outline of operations; which, with so vast a subject to handle makes it difficult already to confine myself to such limits as will make the book concise and cheap enough for every grape grower in the state. But a short outline of what has been done so far, would seem necessary and proper, to show what we may expect of the future, and may well be expected of me.

It is well known that the earliest beginnings were made by the Jesuit fathers at San Gabriel, with what has since become known as the Mission, or as it is erroneously called by many, the California grape. It is no doubt a true *Vinifera*; whether, as some believe, it was grown from the seed or from cuttings imported from Spain, it certainly bears no resemblance to our native wild vine, *Vitis Californica*. A few enterprising men saw in its success there the probabilities of a valuable industry. Their experiments were rewarded with abundant crops which even surpassed their expectations, as our dry and equable summers favored the development of the grapes, and although it was thought in those days imperatively necessary to irrigate the vines, they found that the Mission always ripened its fruit, would produce large crops, under a very simple and convenient system of pruning, and make a fair drinkable wine in most seasons. But when they came to handle the product for wine, they forgot or overlooked that our long, dry summers always give us a grape rich in sugar, and that every fruit has a period in its ripening when it is most lively and most sprightly to the taste. In Europe, where grapes do not ripen so fully, it becomes necessary to let them hang as long as possible, to bring out their full amount of sugar, necessary to make a fine wine; while here they are apt to become over-ripe, and as it needs a certain amount of acids to develop the full bouquet and sprightliness of each variety, the natural consequences of late harvests were very fiery, heady wines; either with a great deal of alcohol, or very often badly fermented, unpalatable and milksour. They were not wines to "make glad the heart of man," but such as would make his head swim and feel uncomfortable. These were placed upon the markets as California hocks and clarets, and did not, as may be expected, please the palates of those who were accustomed to the finer and lighter wines of France and Germany. They pronounced them heady, earthy, and in many cases unfit to drink. The natural consequences of such a course was, that California wines fell into disrepute and could not find buyers at any price; grapes could not be sold at figures to pay for the gathering and working of the vineyards, and hogs were turned in to fatten on their products. This was one of the first mistakes committed; owing partly to an inferior variety of grapes, partly to faulty management of the crop; and retarded for a while the further development of the industry.

But still the incontrovertible fact remained, that some fair wines had been made, that the vineyards produced regularly a good crop of healthy grapes, and that sweet wines could be made, even of the Mission. Grape growing had started in Southern California, and on irrigated land, but it had gradually spread to the more northern parts. Experiments had been made on land without irrigation, and it was found that the vines, though of a slower growth and bearing less, made a more delicate and higher flavored wine than on irrigated ground. Many progressive men, encouraged by the evident success with the Mission grape, imported cuttings of choice varieties for trial from France, the Rhine and Spain, often at heavy expense and risk; they were planted in different sections, and mostly found to succeed well. The introduction of the Zinfandel grape, the first variety from which a creditable claret was made, also gave a new impetus; more care and skill was applied in handling the wines, and they slowly but surely found a market at fairly remunerative prices. Large wineries were built, more improved machinery applied, and the wine makers who had started them, and could sell their wines to the dealers when six months old, at a fair profit on their labor, raised the price of grapes until grape growing became a very lucrative business again. Farmers found that the lands they had cropped with cereals until they were exhausted, and would not produce grain, would still yield large crops of grapes, for which they had a ready market at home. It is certainly not surprising if they became over sanguine, until everybody and his neighbor planted grapes. As the Mission was known to be productive, and they could sell all they could grow, a good many vineyards of this variety were again planted, together with a large acreage of Zinfandel and Malvasia. The vineyards were, to a large extent, planted by men who had little appreciation of fine quality, but planted grapes simply for the money they could make out of them. Rich bottom lands, which were easily cultivated and produced heavy crops, were naturally preferred to the less rich hillsides, with more laborious cultivation and lighter crops. The common system of stool pruning, so convenient and easy, was used for all varieties indiscriminately, and many of the choice varieties, such as the Riesslings, Pinots, and others, did not yield under this treatment, therefore came into disrepute as poor bearers, while with a little more care in staking, tying, and pruning, they would have produced well. Thus the heavy bearers, Mission, Malvasia, Burger, and Zinfandel, were given the preference, even at somewhat lower prices for the grapes, and the planting of really fine varieties followed by comparatively few.

When Chas. A. Wetmore, our past Chief Viticultural Officer, made a trip to Europe, and especially to France, to investigate the resources and methods of those countries, it was but natural that he should be deeply impressed with the magnitude of this, the leading agricultural interest of France, and take the French as models in everything, cultivation of their vineyards, varieties of grapes cultivated, methods of wine making, etc., especially as the resemblance of climates is great in many respects. But he lost sight of the great distance; of our rainless summers, our wet and mild winters, and our immense diversity of soil and climate even in the same vineyard; of our different and more costly labor system, which compels us to look for the cheapest and most simple mode of culture, compatible with thoroughness. While I do not wish to depreciate the great results obtained by the French vineyardists and wine makers, from which we can obtain most valuable information, yet we should consider that it has taken *them* centuries to study the methods best adapted to their wants and surroundings, their soil, climate and varieties, and that we cannot hope to excel here, unless we do the same, and adapt *our* methods to our wants. Practical knowledge, gained here at home, even of a few years, will be a safer guide to us than to blindly follow the practices of a people thousands of miles distant, and who differ just as widely in their application among themselves as we do here. If we try foreign methods, appliances and varieties, let us do so cautiously, thinkingly, and with due regard

to the differences which naturally arise from all accompanying circumstances. I do not think that any one will question
the fact that serious mistakes have already been made by following French methods entirely and blindly, and especially
in supposing that California wines, in their infancy and imperfect state of development, could already compete in the world's
markets with the average of French and German brands, with
their prestige of centuries, their intimate knowledge of the markets and their requirements, the blends that will produce the
most harmonious results, composed of choice varieties; when
we had only the product of a few inferior varieties to offer in
quantities to cut any figure in the markets, while our really
choice samples were hardly seen or known, and then only
to very few.

It was a serious mistake to advise the unlimited planting
of vineyards, and to create the impression that this State
could not produce enough of good, cheap, wholesome clarets
to fill, or ever glut the markets; and those who advised such
a course lost sight of the fact, that, before we can count our
share of the custom of the world, we must not only overcome
the prestige of other nations, but also the prejudice which the
inferiority of many of our earlier productions have created
against us. The vintage of 1884, with its abundant and
rather inferior product, followed by a panic in prices, was a
lesson by which we should profit, as it should have taught us
what we may expect. Had this been followed by an equally
abundant and similar product in 1884, with the large area of
additional young vineyards which came into bearing, what
else could we expect but prices so low that they would hardly
pay the producer? This was averted by the very light crop
of 1885, so that wines came up to fair prices again. But
after the crop of 1886, which was a good one in quality as
well as in quantity, perhaps as good as we can ever expect,
prices have dropped again, and those who planted vineyards

with the idea that the condition of the five preceding years would remain the same, that we could never
produce enough of good, cheap claret to meet the demand,
and have in consequence of it, planted mostly Zinfandel,
often in locations not at all suited to that grape, on soil
which will never produce it in perfection, find themselves
confronted by low prices and slow sales. While it makes a
really *fine* wine in choice localities, and especially on our
hillsides, rich in iron and other minerals ; I have still to see
the first really superior claret made from it on our rich bottom
lands, where it has mostly been planted. Besides, it needs
skill and knowledge in gathering and fermenting its grapes, to
bring out all its best qualities, which many of our wine
makers do not possess, and the time is coming when three-
fourths of our Zinfandel and Mataro wines have to be sold as
inferior, and only one-fourth will be classed as strictly fine
wines, and sold at remunerative prices.

Another mistake was made in discouraging, or at least not
to recommend, the planting of fine white wine varieties,
although we may safely claim that we have more first-class
white wine grapes than red. We can produce choicer white
wines to-day, to suit more different palates, and make a
greater quantity to the acre than of red. Yet the cry has
been: " Red wines are the universal drink;" therefore plant
them, until the public has planted generally about four-fifths
of red to one-fifth of white wine grapes. We see the effects
of this already in the higher price and greater scarcity of
white grapes and wines, which bring one-third more in the
market, and are more sought after than the red. What then
will it be in the future, when the large quantity of red varieties planted will come into bearing ? I grant that there is a
larger quantity of red wine consumed in the world's markets
than of white, but not in the proportion already mentioned;
and I think we will do wise to plant more white varieties in

the future, and perhaps graft some of our young and old red varieties with choice white ones. We cannot deny that there is a large part of the wine consuming public who do, and always will, prefer the more delicate white wines to the more astringent red, and we should try to suit their palates, especially when they are willing to pay better prices for them.

Although wine making is a very simple process in itself, yet it needs great discrimination and judgment, as the product of each vintage is apt to be different, and the temperature of each season is also a very important factor in fermentation. It was evidently judging from the experience of the vintage of 1884, with a product low in sacharine, late in ripening and a cool temperature, that Mr. Wetmore made the assertion, "that any one could ferment his grapes and make his wine in an old shed, and turn out a good drinkable wine, without cellars or costly fermenting rooms." The prevailing idea seemed to be, that the quicker fermentation could be excited, the better would be the wine, even if this had to be accomplished by adding brewers yeast or flour. The season of 1885 came, with entirely different climatic conditions, a different product, an early vintage, and what was the result? Many, who themselves had no practical insight into the principles governing fermentation and wine making, had followed this advice, and put their fermenting tanks out in the hot sun, with no other covering than a few boards, and the temperature over 100 in the shade. Fermentation set in with terrible violence as could be expected the grapes, though rich in sugar, were sluggish and without life from the long continued drought, and the result in a great many cases, were a suddenly checked fermentation with from 2 to 8 per cent. of unfermented sugar, or rather caramel in the wine. Mr. Wetmore was applied to, to help them out of this difficulty, and certainly brought a good deal of energy and good will to this herculean task. Following the advice of French wine makers, who are

also known as the greatest wine *doctors* in the world, he advised in rapid succession, brewers yeast, gypsum, fresh grapes,
fresh ferment from other vats, then tartaric acid and tannin.
But alas, in spite of all these remedies and their application
many of the new wines, being "stuck" once, refused "to go
through" and had to be worked into sweet wines, or distilled
into brandy. If our wine makers have gained in experience,
it has been a *bitter* and costly one, although some of their
wines remained *sweet*, and as wines became scarce, and the
unfortunate practice still prevailed that the dealer had to
buy a whole cellar, and had to take the good with the bad,
these imperfect wines were doctored up, sent to the East
and elsewhere during the season of low freights in 1886, and
the markets flooded with indifferent wines by unscrupulous
persons, which again damaged the reputation of California
wines seriously, until now, their purity, on which we have always justly prided ourselves, is called into question. The
outcome of this was the enactment of the "Pure wine bill" as
it is generally termed, by our legislature last winter, which,
although perhaps susceptible of improvement and amendment,
will at least show to the world at large that we understand
wine to be the *pure unadulterated, fermented juice of the grape*,
the healthiest and best drink for the million. And what inducement can we have to adulterate it? Surely grape juice
pure and simple, is cheaper in our blessed climate, than any
decoctions or sophistications; and we need nothing else, as
soon as we are fully informed about the processes of making
it. To assist in this, is the principal object of this volume.

Another mistake which many of our planters have committed, is the persistence with which they have planted, and are
planting even now, the vinifera cutting and vines, in districts
affected and nearly destroyed by the Phylloxera. They ought
to profit by the lessons taught in France and all over Europe,
by the devastated vineyards which have reduced the crop of

France to about one-third of what it was formerly, until the greatest grape growing nation on the face of globe cannot raise sufficient for her own consumption, and has to buy from all her neighbors to meet the demands of her customers. The devastations already made in our own vineyards would have convinced the most skeptical, that they ought to avail themselves of the only efficacious remedy, the planting of resistant vines, the cheapest, simplest and best preventative. If, instead, they persist in planting vinifera, they may find themselves in the near future with wine cellars and casks, but no grapes to fill them. But perhaps this may also be a blessing in disguise, as it may prevent over production, and take off a great many of the old vineyards of inferior varieties, making room for better kinds.

Another great error and a crying evil at the same time, is the high price at which wine is mostly retailed in this State. Is it fair or prudent even, that wine which can be bought by the barrel at from 25 to 30 cents per gallon, should be sold by the glass, in the majority of our saloons, at 10 cents per glass, and that glass be so small that it will take from 60 to 80 to make a gallon? How can we ever expect to see wine what it ought to be, the daily beverage of our people, enlivening and strengthening them, and making them truly temperate, when it is retailed at such enormous profit, the retailer charging 6 to 8 dollars per gallon, for what costs *him* 30 cents? The same may be said of our hotels and restaurants; the majority not even keeping California wine under its own honest name, but selling it under French or German labels at 75c. to $1 per bottle. But I am glad to say that there are honorable exceptions to the rule, and that some of our hotels and restaurants already serve it on their tables instead of tea or coffee, if the guests prefer it. I know of one hotel even, and that what is called a "second-class" house, where guests are served with a good and plentiful meal at 25 cents, and a bottle of wine is served with each two plates at dinner. This place alone uses 2000 gallons of light wine annually; yet our so called first-class hotels, who charge their guests $3 per day, pretend that they cannot afford it. But the remedy is very simple. Let us leave such houses severely alone, and patronize only those who are willing to do the fair thing towards us, or buy wine by the gallon from the producer, keep it at our homes, and enjoy it with our families.

I have so far reviewed only the wine interest as the leading and most prominent one. But it is far from being the only branch of grape culture followed. Our raisin industry has also assumed large proportions, and though it lagged and suffered under similar disadvantages as the wine industry, being also a new and untried business, with which those who entered into it were mostly unacquainted, yet it seems to have passed its worst period of supression in prices. The growers have learned better methods of curing, use more care and skill in packing, select their fruit and grade it better, so that many brands of California raisins already rank with the best imported goods and bring the same price. Our dry falls greatly favor this business, which bids fair to assume gigantic proportions, and to offer a pleasing and wholesome occupation for women and children, certainly more wholesome and pleasant than the work in crowded factories.

The growing of grapes for table and market is also receiving a new impetus through cheaper Eastern freights and better methods of packing, quicker transportation, and improved shipping facilities. There seems to me nothing to prevent, that California fresh grapes should be in the market from August to February, and even later. Our earliest locations, at Vacaville and Pleasant Valley can furnish ripe grapes in August, while the Santa Cruz mountains furnished them fresh from open vineyard, without the slightest touch of frost, last winter, in January; they can go through to New York in six days, and at moderate charges for freight, where Eastern freights and time of transit formerly were almost prohibitory. In summing up the past and looking at the present, what do we find? A great industry, which forty years ago was hardly thought of; an untried field, over which we have worked and experimented with bright hopes, alternated with discouraging reverses, but which has already brought forth results of which we may feel justly proud, and which ought to encourage us to renewed exertions in the future, We have already produced wines which can safely compete with the best foreign importations, and have the great advantage of being sold at less than half the price. I speak knowingly, for I have had frequent opportunities of testing the best importations, even including the finest Johannisberg, Forster Traminer, Chateau Yquem, and Clos Vougeot; and with only four exceptions have tasted as good and better wine in California. We can produce a good, sound wine every season, and will have a great improvement in its general quality in the near future. That manifold mistakes were made and errors committed was but natural; and I have enumerated some of them not actuated by a spirit of fault finding, but by the conviction that we must know our shortcomings to enable us to do better in the future. We can make all kinds of wines, from the light, pleasant wines of France and Germany to the heavy and fiery ports and sherries of Spain and Portugal, and they have already been introduced in England, Germany, Holland and Belgium, the Sandwich Islands, even Japan and China, Mexico and all the States and Territories. If they have not always given satisfaction, they can be made, and will be made to do so as soon as could possibly be expected. The State has fostered and encouraged the industry, by creating and endowing the State Board of Viticulture; it is well represented at our State University, and thousands of industrious and thinking people have chosen it as their occupation. We can boast already of the largest vineyards and

wineries in the world.   We have the finest and most uniform
climate, the most diversified soil and aspects.   Nature has
designed this to be *the* great Vineland, the France of the new
Continent, where every one can " sit under his own vine and
fig tree."   Be ours the happy task to work out this problem,
and prove worthy of it, profiting by the errors of the past,
with hopes that never flag, of its happy ultimate accom-
plishment.

CHAPTER II.

CLASSIFICATION OF GRAPES.

I shall not attempt elaborate descriptions of all the species
now found by botanists, as they would be of little practical
use to the vineyardist.   Suffice it to say, that the late Dr.
George Engelmann, one of the keenest observers of nature,
found a striking distinction in the seeds, and classified them
into fourteen species, in the following order: 1.   Labrusca
or Northern Fox.   2.   Candicans or Mustangensis.   3.
Carribbea or Caloosa.   4.   Californica.   5.   Monticola or
Mountain Grape.   6.   Arizonica.   7.   Æstivalis or Summer
Grape.   8.   Cinerea or Ashy Winter Grape.   9.   Cordifolia
or Winter Grape.   10.   Palmata or Rubra.   11.   Riparia
or River Grape.   12.   Rupestris, Sugar or Bush Grape.
13.   Vinifera or European Grape.   14.   Rotundifolia, Vulpina or Southern Fox.   The accompanying cut will illustrate
the form of seeds and natural size of them.

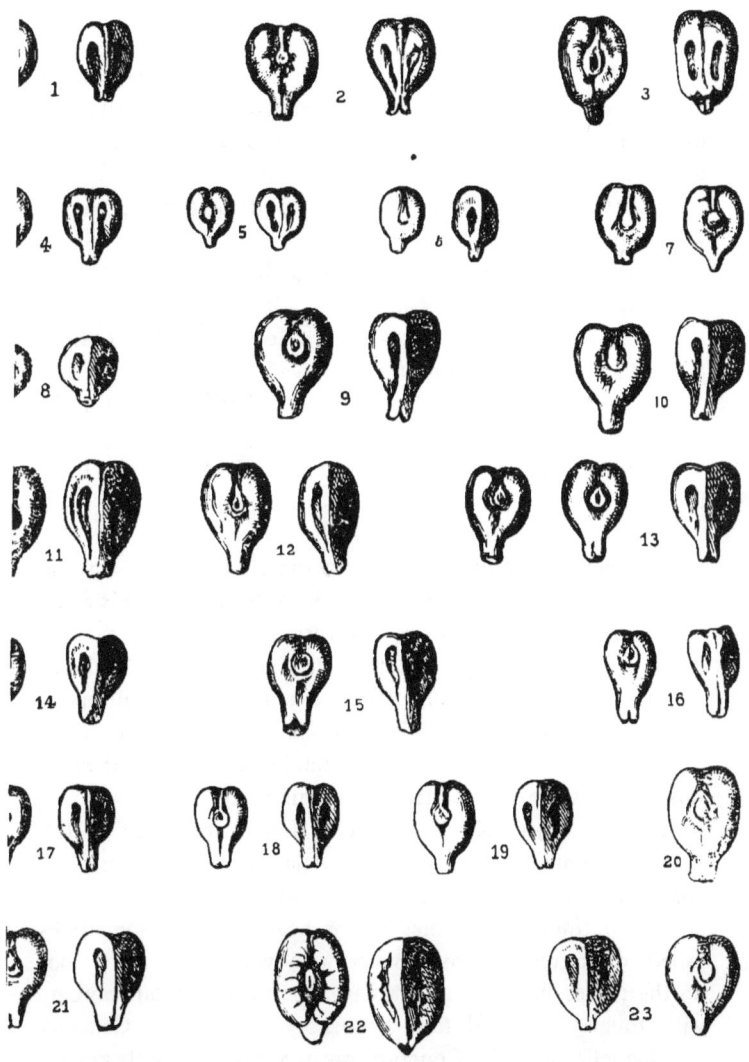

FIG 1.
SEEDS OF CERTAIN AMERICAN AND EUROPEAN VINES.

—V. Æstivalis,
—V. Cordifolia,
—V. Candicans,
—V. Cinerea,
—V. Riparia,
—V. Rupestris,
—V. Labrusca,

10.—Isabella,
11.—Taylor,
12.—Clinton,
13.—Delaware,
14.—V. Vinifera,
15.—Chasselas,
16.—Cabernet,

17.—Jacquez,
18.—Herbemont,
19.—Rulander,
20.—Eumelan,
21.—York-Madeira,
22.—Scuppernong,
23.—V. Solonis.

Mr. T. V. Munson, of Denison, Texas, has lately made another classification according to geographical distribution of the native American species. He classes them in seven groups, as follows:

1. *Riparian Group*, a, Riparia, b, Rupestris, c, Nuovo Mexicana, d, Arizonica.

2. *Cordifolian Group*, a, Cordifolia, b, Palmate.

3. *Cinerean Group*, a, Cinerea, b, Monticola.

4. *Æstivalian Group*, a, Northern form, b, Southeastern form, c, Southwestern form:

5. *Vulpina Group*, a, Labrusca, b, Carribbea, c, Candicans.

6. *Meaty fruited, soft rooted group*, a, California, b, Vinifera.

7. *Rotundifolia* or Southern Fox.

Of these, we are only more immediately interested in the following, which have either been introduced into this State or found wild ; these are, 1, Labrusca, 2, Riparia, 3, Æstivalis, 4, Rupestris, 5, Arizonica, 6, Californica, 7, Vinifera.

The first six are chiefly valuable as stocks to graft upon, though some of the varieties may prove valuable for their fruit; while from the last come all our leading grapes for wine, raisins and table now cultivated in this State.

1. *Labrusca*, or Northern Fox, is found wild east of the Rocky Mountains, mostly on the Atlantic shore, from Canada to the Gulf, generally in moist woods or thickets. Leaves, large and thick, sometimes entire heart shaped, sometimes lobed, dark green above, covered with whitish or rusty wool on the under side; berries, rather large, purple or dark amber, with tough pulp and foxy or musky odor. The Catawba and Isabella, also the Concord, are the most generally known cultivated varieties of this class, and are occasionally grown for market in this State. Its roots are tough and wiry, and have a tendency to run along the surface, which hardly fits

the class for our dry soils and summers. Only partly resistant to phylloxera, although more so than Vinifera.

2. *Riparia*, Riverside Grape. In its wild state we distinguish two distinct forms, the smooth leaved, and downy leaved or pubescent. Its homes are the bottoms of larger streams, especially of Missouri, Illinois, Kansas and Nebraska, where the smooth leaved form is most common, while the downy leaved appears mostly in Texas and New Mexico. Both are equally valuable as grafting stocks. Wood, thin and long, long jointed; the leaves heart shaped, with acute points, deeply serrated and sometimes lobed; in the smooth variety light green above and below, without down ; in the pubescent, wooly below, and the stems covered with light hair. Berry, small, black, without pulp, and sprightly, dark colored juice. Roots, thin and wiry, hard, spreading; seems to succeed in nearly all soils; propogates readily from cuttings. The wild vines are entirely resistant. The most known cultivated varieties are Clinton, Elvira, Taylor and others, some of them evidently hybrids of Labrusca and Riparia.

3. *Æstivalis* or Summer Grape. Vine a strong grower, healthy and hardy, but difficult to propogate from cuttings. Leaves, large, thick, downy beneath, generally lobed, though some varieties are not. Found mostly on uplands, and is eminently fitted to withstand drought, as it has strong, very hard roots, which strike deep into the soil. A fine stock for grafting the Vinifera, as they take readily. Berries, small, black with blue bloom, not pulpy, and some of fine quality for wine. The Herbemont, Lenoir, Rulander and Cunningham are the most prominent cultivated varieties.

4. *Rupestris*. Bush or sugar grape, Southwest Missouri, Arkansas and Texas. Vine bushy, with many small branches, stocky; leaves small and shining above and below, heart shaped; berry small, black, with blue bloom, propagates

readily from cuttings; vine makes a good grafting stock, but is apt to sucker; roots thin and wiry, resistant.

5. *Arizonica.*   Arizona and New Mexico.   Resembles Rupestris very much, but is a more upright grower.   Does not seem to take the graft readily, and has not so far full-filled the expectations of its disseminators.

6. *Californica.*   Generally found wild along all the streams of this State, where it often attains very large size.   Wood grayish, long jointed, a strong grower, with thick, fleshy, soft roots, which go straight down.   Leaves heart shaped, downy and wooly; berry small, black, without pulp, but very large seeds.   It takes the graft readily, and, should it prove' entirely resistant, which is not fully proven yet, will make a valuable stock on deep, moist soils, while not so well adapted to dry hillsides.

7. *Vinifera.*   To this class belong all of our leading varie-ties.   It is the old European or Asiatic grape.   Too well known here to need any more minute description.

This much seemed to be necessary to give the reader a clear understanding and avoid repitition and minute descrip-tion.   Those who wish to investigate this subject further will find the essay by Mr. T. V. Munson, on Native Grapes of the United States, read before the American Horticultural Society, very interesting, which can be obtained from the Sec-retary, Prof. H. W. Ragan.

CHAPTER III.

PROPAGATION OF THE VINE.

1. By Seeds.

This may be divided into two separate parts, according to the object the propagator has in view, namely:

1. To raise new and improved varieties.

While the raising of vines from seed with this object in view is more a labor of love than of actual profit, its influence on grape culture has been so great, and we are already so largely indebted to its zealous followers, that it can not be entirely omitted here. All our fine varieties are either accidental or carefully hybridized crossed seedlings; and there would be no improvement in varieties without this. The immense progress in American varieties within the last forty years, when only half a dozen varieties were known, of which the Catawba and Isabella may be considered as fair samples, are due to the labors of such men as Rogers, Wylie, Campbell, Ricketts, Miller, Rommel and Munson, who have originated varities for their climate and purposes more valuable than our Viniferas would be, and it certainly required a long line of improved seedlings to make up the long list of excellent varities of Vinifera we now cultivate here and in Europe.

To begin then at the beginning; choose your seed from a good stock. Take a good variety which you would like to improve in a certain quality, be it size or form of berry or bunch, fruitfulness, time of ripening, or flavor. If a vine stands next to the one you take the seeds from, which has the desired quality, and which may have impregnated the bloom, so much the better, your chances are so much more.

Choose the finest bunch, the most perfect berries, and either
take the seeds from them fresh, or keep them over winter
in their pulp or, if cleaned, keep them in sand in winter until
they can be sown in early spring. They will not germinate
so readily if allowed to become dry. Make a bed of finely
pulverized soil, the deeper the better; sow in drills about one
foot apart, and the seeds about an inch apart in the rows,
covering about an inch deep with finely pulverized soil, pressing it lightly to the seeds, either with your foot or the back of
the hoe. When the young plants appear, which will generally be within six weeks, keep them clean and well cultivated
through the summer; in the fall, take them up carefully so as
not to mutilate their roots, and heel them in well-drained,
fine soil; covering up nearly to the top to keep them during
the winter and preserve their roots in the best condition. It
will be well, during the summer, to look over them frequently,
and if any of them show signs of disease in leaf or growth,
or are puny or sickly, pull them up, as they will not be
worth keeping. It may also be well to shade the young
plants for the first month or so, to prevent the sun from
scalding them while yet tender; and if any of them grow
very strong, give them small sticks for support. In the following spring they may be transplanted to their permanent
location in vineyard or garden. The ground should be moderately light and rich, and loosened, if at all tenacious, to the
depth of eighteen inches.

Make a slanting hole with the spade about a foot deep,
then shorten the young growth on the vine to about six inches
above the collar, (the part of the vine where the growth from
the root begins). Then spread the side fibres well, letting
the top or leading root go down to the bottom of the hole,
and set the vine about an inch deeper than it stood in the
nursery. Fill up with well pulverized soil, pressing it lightly
with the foot. They may be planted the usual distance

apart in the vineyard, and when the young growth appears, leave but one or two of the stockiest and strongest shoots. Allow all the laterals to grow on these, as this will make them short jointed and stocky, Cultivate well and frequently, keeping the soil loose and mellow.

This second season the young seedlings ought to make a growth of a few feet of short jointed wood. This should be cut down to three or four buds the next spring. These, if the growth is strong enough to develop fruit buds, will generally show fruit, or if any of them look very promising in leaf or growth, fruit may be obtained sooner by grafting the wood on stronger vines. The first fruit is generally imperfect, and will increase in size of bunch and berry for several years. If the quality is good, and they show a fair degree of fruitfulness, they may be considered promising, although it will generally take several seasons to develop them fully.

Quite a number will prove barren, or not of desirable quality. These can be grafted afterwards with the most promising or some good old variety; therefore there is really but little loss, while a lot of seedlings are always a very interesting study, which may be very valuable in its results.

2. To raise seedlings as stocks for grafting.

Here we have an entirely different object in view, and our aim is simply to raise the most uniformly healthy plants, of the strongest and most even growth. The wild species of our vines are more apt to produce these than the cultivated varieties; therefore the seeds of them are preferred. Whether these had better be Riparia, California, or Æstivalis, we will consider in the chapter on " Resistant Vines."

A pound of seed of these wild species will generally produce, if good and fresh, from 2,000 to 3,000 plants, and as it can now be had cheap from reliable men, who make it their business to gather it in its native region, it affords a very cheap and convenient way to raise good stocks, as the price is but from one to two dollars per pound. Prepare a piece of good soil thoroughly, deeply plowing and cultivating it, leveling off well with harrow and clod crusher. It will be more economical to work these with a horse cultivator during the summer; therefore the drills can be made three feet apart. The seed, which generally is dry when received, even if gathered the foregoing fall, should be prepared about a week before sowing, by soaking in hot, but not boiling, water, in which it may remain for twelve hours, when the cold water may be poured off, and another application of hot water given. The next day pour it into a sack to drain off the water, and lay it in the sun during the day, moistening the sack whenever it becomes dry, and keeping it under cover at night. The best time for sowing in this State is in February or March; in frosty locations it may be well to wait a little later. It generally takes about three weeks to a month before the young plants appear, and all danger of frost should be passed then.

Sow in shallow, broad drills, so that the seeds are at least an inch apart, if you wish to raise good stocky plants; cover about an inch deep, with fine soil, pressing the ground to the seeds. The young plants should be kept clean and well cultivated, a shovel cultivator to "straddle the rows," so that one share goes on each side of the row, and run through them every week or ten days, will keep the soil loose and mellow, although they should also be hoed once or twice. In the fall or early in winter, when the rains have softened the ground, they can be dug by running a tree or grape vine digger as the nurserymen use them, under the rows and pulling up the young vines by hand; or if such a tool is not to be had, a furrow plowed away from them on each side, and lifting them with the spade. I sort and grade them generally in two classes, as it will make a more even plantation, tie them in bundles of 100 each, and "heel them in," in beds slightly raised above the surface, when they are ready for planting in vineyard, treating them as described before. In one or two years, they are ready for grafting, and generally make very fine and even stocks, as they can be transplanted with nearly their entire roots and therefore receive very little check. While planting, however, the roots should be kept in a pail of water, to keep them moist and fresh.

CHAPTER IV.

PROPAGATION BY CUTTINGS IN OPEN AIR.

I am aware that a great majority of California vineyards are planted with cuttings planted directly in the vineyard. While this makes very good vineyards sometimes, if the circumstances are favorable, yet I do not recommend it, and think it the cause of the great number of uneven stands we see in the State. Moreover, all varieties do not root equally well, and it is always safer to plant in nursery, and remove the plants into vineyard next spring.

Most of the Vinifera varieties root readily from cuttings, but as I do not advise planting this, on account of the danger from phylloxera, and as nearly all American species do not root quite so readily, it is all the more prudent to plant in nursery first.

Of the American species, the varieties I would chiefly recommend for stocks, are the wild Riparia and the Æstivalis, for reasons which will be given in the chapter on "Resistant Vines." The wild Riparia roots readily, while of the

Æstivalis class only the Rulander and the Cunningham root well, the Herbemont and Lenoir will turn out about 50 per cent, and the Nortons Virginia and Cynthiana hardly any.

The cuttings can be made any time after the leaves have dropped until the buds begin to swell, it is best, however, to make them in the first part of winter, when they can be heeled in bundles, to keep until planted. Neither the very large and pithy, nor the very small wood near the ends should be selected, but rather the medium sized, short-jointed wood, which will not only be more sure to root, but also make a firmer and better plant. Nor is it advisable to make the cuttings 18 inches or even two feet long, as has been the practice in this State, very likely derived from the old European vintners, who follow that practice for no other reason than because their father and grandfather did so before them. Forty years of nursery practice, commencing with 18 inch cuttings, and ending with 9 to 10 inches, has taught me that the longer the cutting, the more feeble and small will be the roots they make ; small puny fibres, distributed over the whole length, instead of the strong, well developed root system at the base of the shorter cutting. If the lower end of the cutting or plant is buried in the cold hard soil, below the influence of sun and air, so necessary to all plant life, how can we expect it to make strong, healthy roots? Here in California it may be well not to go into extremes, but a cutting of 12 inches, from the lower bud to the upper, is long enough, and will make a better vine than 18 inches. That the wood should be well ripened and sound, is, of course, the first consideration.

The vines may be pruned in fair weather, and the clippings taken in to be worked up during rainy days, but it should never be allowed to get dry, as that destroys its vitality. Cut close below the lower bud, making the cut somewhat slanting, as the accompanying illustration will show, leaving about an inch

of wood above the upper bud
or eye. If a small piece of
the old wood, or the whorl of
buds, where it starts from the
old wood, can be left so much
the better, such cuttings are
almost sure to grow. They
are then tied in bundles of
250 each, the lower ends
made even. I use leather

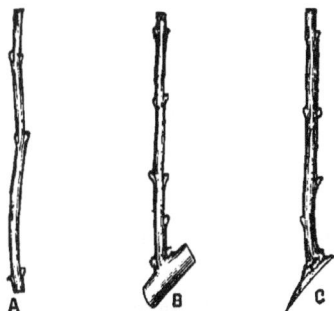

FIG 2.

A, ordinary cutting; B, cutting with
old wood (mallet cutting); C, cutting
with longitudinal sections of old wood.

straps and buckles to draw
them together and then tie
firmly with annealed wire No. 16. This is a much better tie
than twine or bale rope, as it will not rot, and is much cheaper besides. The bundles are then "heeled in" or buried in
the ground, in trenches made for that purpose ; inverted, that
is placed on their tops, and the butts well covered with 3 to
6 inches of soil. Inverting them has the object to place the
lower end of the cutting, which is to form the callus or roots,
closer to warmth and air than the top buds. It will thus callus first, while the top buds remain dormant, and is ready to
throw out rootlets as soon as planted, while with the tops
above, they often start to grow before there is anything to
support them, and then wilt down afterwards.

In the spring I wait until the ground is warm enough, generally until April in Northern California. Select a piece of
good, deep soil for the Nursery, which should be made mellow
and friable by repeated deep plowing, if not naturally so.
Then throw out slanting trenches with the plow, deepening
them with the spade if necessary, three or four feet apart.
Put in the cutting as close as convenient, say two inches
apart in the row, slanting them enough so that the lower end
is 9 inches below the surface, while the upper bud is about

even with it. Draw in fine soil with the hoe, and firm it well around the base of the cutting, as it is important that it should be closely packed around it, filling up to the top of the cuttings, and if some fine soil is drawn over it, it is all the better for this mulch. Keep clean with hoe and cultivator, and the soil stirred frequently during the summer, to keep it mellow and moist. I have found this much better than irrigating, which makes the ground cold and hard, during a time when the young plant needs warmth as well as moisture. If the ground is well stirred, it will never dry out deep enough to injure the roots. The plants can then be taken up in fall or early winter, and handled and planted similar to seedlings. If for particular reasons, it should become necessary to plant cuttings immediately in the vineyard, I would advise to plant two, instead of one, making a hole with a spade, and getting the base of the cuttings about six inches apart, so that, if both should grow, one can be removed and planted elsewhere. In this manner, we can avoid vacancies, as either one or the other will usually grow.

CHAPTER V.

PROPAGATION BY LAYERS.

This is but little practiced in California, owing to the facility by which all the *Vinifera* species can be propagated by cuttings. It is, however, valuable in propagating such varieties as have very hard wood and will not root readily from cuttings, especially of the *Æstivalis* class; and for filling vacancies in old vineyards. I will first describe the process which we call surface layering, for the purpose of raising a large number of young plants from hard wood varieties.

Choose a young cane of last seasons growth, starting as near to the base of the vine as possible. It will be well, the summer before, to leave some of the lowest shoots for that purpose, growing them as long as possible. This cane is pruned as long as it has well developed, sound buds; the ground made mellow below the vine, and a shallow trench, say two inches deep is drawn with the hoe as long as the cane. This is then bent down into the trench, and fastened on the bottom with small pegs or wooden hooks. (Fig. 3.) Each bud on the

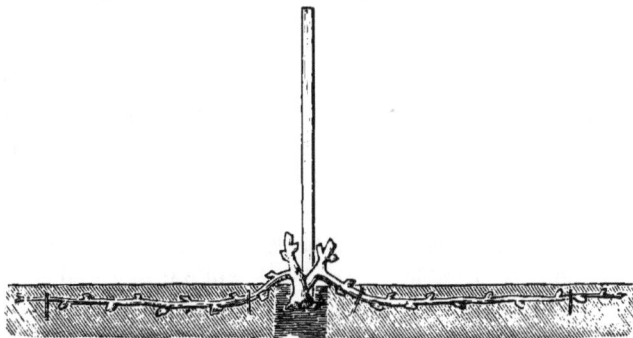

FIG. 3.

cane will generally produce a shoot, which will grow upwards, or should be made to do so, when the trench is filled up around the shoots, which is done when they are about a foot high. Each of these shoots will then throw out roots around its base, and in fall or winter they are dug, beginning at the furthest end of the cane, cutting the roots with the spade at proper length; the plants are divided by cutting behind each shoot with the pruning shears, when each will have its own system of roots, the shoot making the stem of the vine, which can be shortened in at planting to the proper length, or this can be done when they are dug. (Fig. 4.) They make

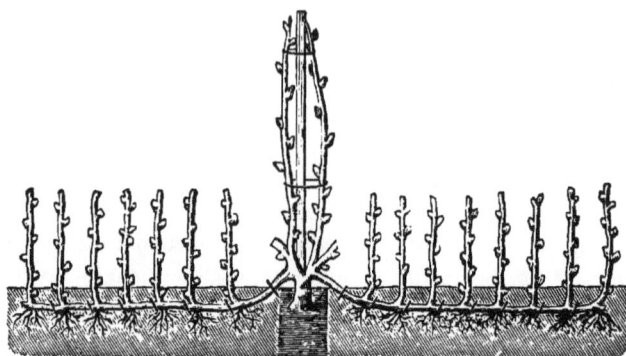

Fig. 4.

very good, strong plants.   The same, or a similar process may
also be followed on young growing canes in summer (summer
layering). . These will not make quite as strong plants as
spring layering, but has the advantage that it can be done
after the plowing and hoeing in spring has been finished, and
therefore does not hinder from cultivating both ways, which
layering in Spring will not permit.   For this purpose, strong
growing young shoots should be left as near the surface of
the ground as possible, as the layers will not root well when
they must be bent down, and afterwards raised again, but
should remain as near the horizontal position as possible.
They are therefore left to trail along the ground and in the
middle of June (here in Northern California,) their leader or
end is pinched off, so that the laterals will grow more vigorously.   They are then laid in shallow trenches, about two
inches deep, and covered with finely pulverized earth.   The
leaves opposite the laterals on the main shoot may be taken
off for greater convenience, also to pack the ground more
closely around the laterals.   When the ground is filled up
around them, they should have a watering, which will assist
greatly in the formation of roots, and the laterals raised as
much as possible to a horizontal position.   For fall, they are

divided up, as each lateral will generally make its own system of roots, similar as in spring layering.

Layering to fill vacancies in old vineyards differs in so far, as a trench is dug from the parent vine to the vacant place, a young cane, grown for that purpose is then pruned long enough to reach to the vacant spot, but the trench must be dug enough to be below the reach of the plow, say 10 inches. The cane is then bent down sharply at the mother vine into the trench; laid on its bottom, and bent nearly at an angle at the vacancy, where one or two buds are left above the ground. (Fig. 5.) The trench is then filled up, and as the layer

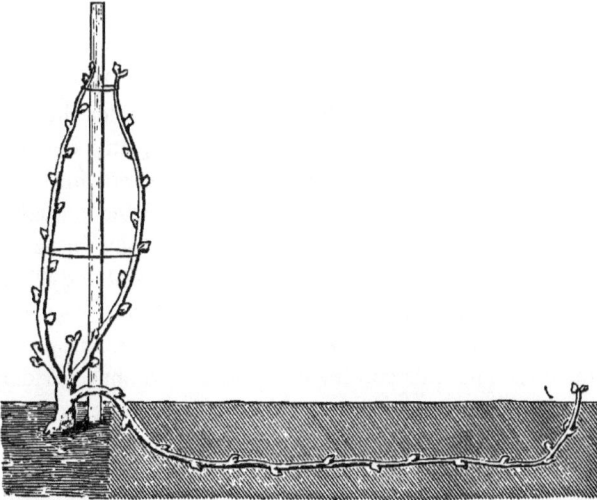

FIG. 5.

draws nourishment from the parent vine, and also forms roots at every joint below ground, it makes a very strong growth, and is able to bear the second season, when it is generally detached from the mother plant with the pruning shears. Every one knows who has ever tried it, that it is almost impossible to fill vacancies with young plants, when the vineyard has come to bearing size. Then this comes in as the only remedy. It is easily perceived, however, that this is also a tax on the mother vine, and allowance must be made for it in pruning.

CHAPTER VI.

THE PHYLLOXERA QUESTION.

That this is a serious one, likely to effect our industry in all its branches, will hardly be denied by anyone. If we look at the devastated vineyards of Europe, if we consider the ruin it has brought to thousands of formerly happy and contented homes in France, how its ravages have decimated this leading industry, so that now they do not produce wine enough for their own consumption, but must buy where they formerly almost supplied the world; how its ravages are already felt in Algiers, in Austria and wherever vines are grown,— we will hardly question that it is *the* great disaster threatening everywhere, including this continent. Indeed, we have evidence sufficient of its destructiveness in this State, to convince us that it is the most formidable enemy of our industry which we have to encounter. It is worse than useless to try to ignore it, as has been done in some sections of the State, it will make itself seen and felt, and no mechanical or chemical means have as yet been found that are of real practical value. All the insecticides that have so far been tried, have proved too costly and impractical in their application; and we must resort at last to the only practical preventative, now recognized by all nations to be their salvation, viz., "American resistant vines."

But while the phylloxera is the greatest enemy to the grape-vine, yet its ravages to a certain extent may be a "blessing in disguise," especially for us here, where there is already the cry of over-production. It may to a certain extent prevent this, and has already decreased the production very seriously in certain sections of the State. It will naturally destroy a large amount of Mission, Malvasia, and others of the old, in-different varieties, helping to take their wines out of the mar-ket, and making room for a better product from choice vari-eties, grafted on the young vines of American stock, and make a better reputation and prices for our vines. It may kill out the vines on many locations not suited to them, and thus improve our coming product indirectly in many ways. If we plant American vines on soils really adapted to them, we need have no fears of the ultimate result. As the first step to successfully fight the enemy is to know where and how to find him, I have quoted liberally from the report of Prof. F. W. Morse, whose close study of the insect and its habits at Berkeley and elsewhere have made him entirely fa-miliar with it in California, and as it differs somewhat in its habits here from those observed by European authorities, they will be of greater value to us than theirs.

1. OBSERVATIONS ON THE LIFE HISTORY AND HABITS OF THE PHYLLOXERA IN CALIFORNIA.

Made from 1881 to 1886 by F. W. Morse, Assistant in the General Agricultural Laboratory.

In the following pages I give a summary of the results of observations made upon the phylloxera, since its discovery in the University vineyard plot in November of 1881. Partial reports of the same have already appeared in previous publications of the Agricultural Department, the whole of which, with additional observations made since, are here arranged in a somewhat modified form.

The observation was taken up immediately after the discovery of the insect, the object at first being simply to study the more prominent types here, in order to become familiar with the prevailing forms generally known to exist in other vineyards; and, also, to note any special habits wherein· they might differ from those observed in other countries. No special attention was originally intended to be given to the prosecution of new investigations tending toward the settlement of disputed points regarding the biology of the insect.

No stated time was set apart for this work, which has at all times been carried along in conjunction with other University duties. The available apparatus, too, has not always been all that could be desired for obtaining thorough and complete results.

At the outset, only one form of the insect was recognized, but as the work advanced, new and unfamiliar forms continually appeared, until we have, by a happy coincidence of special fitness of vine varieties and surroundings, witnessed the production of most of the forms known to foreign investigators.

The importance of such a line of investigation, under such circumstances, becomes apparent, when we consider that among the various forms which the insect is capable of assuming only a part, and these of the forms which are least easily spread, have thus far been seen in appreciable numbers in California vineyards. A solution of this apparent divergence from the habits of the insect, as observed elsewhere, gives direction to the investigations which have heen carried on.

THE UNIVERSITY VINEYARD PLOT.

The University vineyard plot, in which the field observations were made, and the specimens for laboratory work were taken, is situated upon the north side of the University grounds, and upon the extreme east of that part of the tract which was placed under the control of the Agricultural Department, and set apart for experimental purposes. It has a southeast exposure, which renders it somewhat more favorable than the lower lands to the earlier " putting out " of the vine and consequent longer season which is needed in this locality.

The soil is a heavy, refractory clay loam, not easily cultivated except upon the surface, and is underlaid by a stiffer clay at a depth of a few inches, followed at a slightly lower depth by an intermixture of coarse gravel and rocks, thus forming a soil ill adapted to vine growth, and a sub-soil not easily penetrated by the roots. The plot, which is from the nature of the soil difficult to drain, is relieved of this trouble, to a great extent, by the steep inclination and loose-walled bank on the lower side.

The total number of old vines (and some young ones intermixed, which have been grafted on resistant stocks) amount to only sixty-eight, including, beside the common *vinifera* varieties, some that in other countries have shown special fitness for certain forms of the phylloxera, and which are not found among the infected vineyards of our State. Some of these are hybrids of stock, elsewhere bearing the gall type of insects in abundance.

Thus we have been specially fortunate in having a collection of varieties which, presumably, make it possible to produce and study all types that have been observed elsewhere.

The Winter State of the Insect.—The first important point to be considered was to determine the habits and movements of the insect at different seasons of the year, and especially to note their condition during the winter. This was accomplished by repeated examinations during each of the winter months of vines most productive of the insect. No obvious movement to lower parts of the root-system, such as was predicted, was noticeable; but merely a dying out of the different forms upon the older and most decayed parts of the root; the healthier fibres and wood always supporting the insect most abundantly. Even in mid-winter the wingless root form appeared clear to the surface, and even above the ground, where tuberous spots afforded them nourishment and protection. If the winter proved too severe, portions of the root-bark were often found covered with black "lumps" of dead phylloxera mingled with small and shriveled brown ones, in positions similar to those in which they were found late in the fall.

FIELD OBSERVATIONS DURING THE SEASON OF 1881-2.— The observations during the winter of 1881-2, which was cold and was followed by a late spring, revealed the insects numerously settled upon all parts of the roots, and even in mid-winter at the surface of the ground on the base of a last year's sucker. On December eighth the insect was found in decreasing numbers on the old roots, and confined to groups of five or six on the healthiest parts of young roots. They were mostly of the mother form and only a few larvæ. Later in December the preponderance of the mother form was still apparent, and the general settled conditions of the groups was specially noted.

Frequent examinations during the months of January, February, and March, showed no special change, save a more sluggish condition and a darkening of color. •

It was not until the twenty-eighth of April that signs of returning life were visible. Scattering eggs, in groups, from three to five in number, very transparent and quite large, were found surrounding some of the more mature insects which were then becoming yellow but still seem not to have moved from their winter positions. No very young larvæ were to be found, although they were soon after, May first, hatched from a bottled specimen taken from the vineyard at this time and placed under more favorable conditions, in a warm laboratory. A casual examination of specimens put aside in an unsealed fruit-jar proved the roots to be literally covered with insects, only a few of which were mothers; the remainder consisted of young larvæ and eggs. The warm and even temperature of the room was undoubtedly the cause of their earlier activity, and demonstrates the fact that a relatively high temperature is all that is required for the continuance of activity during the winter. This presumption is further borne out by their earlier appearance in 1884, when the spring was fully six weeks in advance of an average year. In the early part of March, before the beginning of the budding of the vine, specimens could be found which would do credit to the insect under the favorable conditions of fall.

The month of May showed a slow but steady increase of the different forms, beginning particularly under the bark of older roots, and later increasing most rapidly on fibrous roots of the present year's growth. The insects seemed to be of a more greenish color than at other times. A marked increase in the rapidity of production was noticeable during the last days of June; still the spread had not yet become general for the season.

First Appearance of the Winged Form.—It was in the following month (July, 1882) that the first indication of the winged form appeared. Insects resembling larvæ were found, with black antennæ and legs, and upon each side of the back, extending along the body, were dark spots, covering the rudimentary wings which distinguishes this as the wing-pad, or "pupa" stage of development. One of these insects was preserved, and by the twenty-sixth of July had developed into a small, live, fertile winged phylloxera, and was transferred to a small vial, where she laid a single egg and died. Others of these pupæ were found later without much difficulty on similar roots ; and others of the winged form have since been easily developed. Their changes during metamorphosis into complete forms were watched with exceeding interest, but need not be described here.

Some of the insects which were most developed, and had the wing-pads well formed, were placed upon a glass slide, covered with a watch glass, and then properly arranged under a microscope. The transformation soon took place. A shedding of the skin precedes the spreading of the wings, and is begun by the dark skin separating horizontally over the wing-pads, a part shedding toward the abdominal segment, and the remainder passing over the head and legs, thereby changing these members and the antennæ to a much lighter color. This operation, in one case, was completed in about fifteen minutes. The wings, which appeared to be folded in a light colored bunch directly across the back, now part in the middle, giving them the appearance of two white or light-colored sacks. These gradually expand laterally from the base, continually carrying the apparent casings in a bunch at the end of the wings until the full length is reached, after which the greater part of the spreading is done. The whole apparent casings are merely folded wings.

One hour and fifteen minutes was consumed in passing from the beginning of the shedding of the skin to the complete winged insect, which soon became active and began crawling about. The body of the insect, in the meantime, had passed from a light hue to a more intense dark yellow, or golden color, while the wings, which before the moulting were black, pass to a light or white opaque, then become transparent, and, as they spread, become thinner and darker, and in the fully developed condition are almost black.

A newly developed winged insect was taken from a moist bottle and placed upon a glass plate, where she soon made attempts to fly. Her wet wings were repeatedly brought perpendicularly over her back and rubbed together, apparently to free them from moisture, and then she attempted to leap, or fly, often raising, by a peculiar curve, from two to six inches high. With the last attempt she flew away.

Roots Producing the Winged Form.— The vigorous, bushy, fibrous growth, or network of roots around the tap-root, evidently caused by manuring and moisture, seemed best to fulfill the necessary conditions for the development of the winged females. It was later shown that these fibrous roots were specially productive of this form; in fact, it is very seldom that it is found on other roots.

Conditions of their Development—Invasion of 1884.—Here it may be well to suggest as a possible explanation for the greater production of the winged form in France, that the more thorough fertilization of the vineyards in that country has favored a more general growth of the surface roots upon which the winged form is mainly produced ; also, that the late spring and summer rains bring about a similar condition of growth. The summer rains alone of France, which last far into *August and September*, would be quite sufficient to produce surface roots of the kind required to produce winged insects in great abundance. It is just previous to this time that the winged invasion occurs, when the insect is carried in swarms to adjacent vineyards. This is a point which I believe has never been suggested, and the truth of which seems to be supported by experience in this State, for neither of the conditions spoken of is ordinarily realized to any extent in California. But the unusual summer rains that occurred in 1884, and which were followed by such a great general development of insect pests in 1885, seemed to impart also to that of the winged phylloxera, a similar impetus ;

as for the first time since observations were begun, this form was during that season found to issue in swarms like those observed elsewhere, doubtless greatly increasing its spread.

We are further justified in supposing that there must be some peculiar condition for their development in districts where only occasionally evident signs of their workings are visible; and in no case have the winged insects been found in such numbers as were developed in 1884.

At the middle of October, 1882, the insects were still numerous, eggs plentiful, and the mothers still laying; the winged form had entirely disappeared. About the middle of November I found only one of the mother-lice fully developed; the remaining insects scattered over the roots were young larvæ, healthy and quite active. Some were blackened and lifeless, but still retained the larval form. Only one egg was found.

December twentieth none of the mother insects could be found; only larvæ, bright but motionless, were present.

SUMMARY OF OBSERVATIONS IN 1881—82.

The condition of the insect during different times of the year may therefore be summarized as follows; There is a dull, lifeless condition of both larvæ and mother lice during the winter, lasting until about the middle of April of a late season, when the hibernating mothers begin to lay their eggs. The young larvæ soon begin hatching out and scatter to all parts of the roots. The increase is very slow until the middle of June. The winged form begins developing about the first of July. Eggs are most numerous about the last of July or first part of August. The old mother lice are soon found in decreasing numbers, and young larvæ are most abundant. A gradual decrease in the number of insects begins about the first of October. No eggs, or scarcely any, are to be found after this month. Very little action or life is noticeable after November.

OBSERVATIONS IN THE LABORATORY, FROM 1881–1885.— For the better verification of the facts observed in the field, and especially for the observation of the habits of the insects during propagation, a series of laboratory experiments was carried on simultaneously with the field observations.

Conservation of Root Specimens.—The first difficulty met with was to preserve the root specimen, in such a shape that it would not mould, and yet remain sufficiently moist to afford sustenance to the insects. This is best accomplished by placing a piece of root containing the required number of insects into a wide-mouthed bottle, supplied with a close-fitting cork. If it becomes necessary to remove the cork very often, a few drops of water may be dropped into the bottom of the bottle to supply any lost moisture, and then, by regulating the temperature, the water can be vaporized and condensed so as to reach all parts of the root. Some specimens were kept in the sunlight, with a good circulation of air through the bottle, but the insect did not thrive under this treatment. Roots thus treated are more difficult to keep in good condition, and the insect becomes more restless. A cool dark place seems best fitted for these experiments.

The leading questions studied were :

1. Number of eggs laid by each mother louse.
2. Rate of laying.
3. Time required to hatch them into larvæ.
4. Time from the hatching of larvæ to the egg-laying age.
5. Pupa form.
6. Winged form.

Number of Eggs Laid.—Two or three specimens containing isolated mothers were placed in bottles and observed every few days. The highest number of eggs from any of these insects was about seventy-five. Numerous bottled roots have specimens of sixty to seventy eggs and larvæ together. It is not uncommon to see a nest of forty to fifty in a row, upon one end of which the young larvæ are just hatching out and moving away, while at the other end are newly laid eggs and the old mother, now reduced to a very small, dark colored ball without apparent life or insect shape. This large number is not so frequently found in the vineyards.

The most prolific insects do not seem to produce a generation of the numerous egg-laying kind. As soon as the eggs are hatched into larvæ the latter move away, while those insects producing eggs that are destined to become laying mothers lay but few eggs, which when hatched move less rapidly than the other kind, and are often found in groups.

Rate of Laying and Time of Hatching.—To determine the rate of laying, properly isolated individuals were watched during their complete season of laying. The rate was found to be very irregular. depending largely upon circumstances. It often amounted to five per day, while at other times the insects ceased laying altogether for several days. A single individual laid thirty-five in seven days ; another thirty-four in two weeks. At the end of four weeks the whole generation had left the spot. The relatively limited supply of sap furnished by a detached root probably served to restrict the number of eggs laid. Still another insect, under less favorable circumstances, surrounded by a meniscus of water, which kept her almost submerged, continued laying for a much longer time and at a much slower rate. It was found that it required about thirteen days to hatch the eggs.

Duration of Larval Condition.—By deducting the time for hatching from the total number of days from the first egg-laying to the egg-producing period of the second generation, we have the time of the larval condition. This was found to be about seventeen days.

Much interest attaches to this form, since it is through it that we know the conditions which will produce the winged form. The later is found with great difficulty upon the vine, while the former is readily distinguished among its associates, the common root louse or larva, which is first sought after in the examination of an infested vineyard. The pupæ once found, it becomes easy to trace them to the winged form, which is usually near by.

Conditions Governing the Production of Pupæ.—My first observations on the fertile winged form were made on specimens accidently produced in the laboratory. By tracing back to its origin the root upon which these first individuals were found, the needful conditions and surroundings could be determined, and thus the winged form could be sought for more intelligently, and found in greater numbers. As before noted, the small, soft, fibrous rootlets of the current year's growth had seemed to be most productive of the form from which the winged insect is developed.

Movements and Transformation of the Pupa Form.—The movements of pupa, or wing-pad insect, observed chiefly during 1884, were not found to be altogether along the smaller roots and thence up the main body to the crown of the root, as is usually supposed, but the insects frequently left the roots and passed up through the soil, which in no place was less than three inches deep. This movement afforded an excellent opportunity for determining where the transformation into the winged form takes place.

The insect in various stages of development could be found in the earth from the surface to the roots, the most incomplete forms being found deepest below the surface of the ground. Some were found under stones, and in such positions as to place it beyond a doubt that they passed through the changes underground, and came to the surface in a trans-

formed condition, contrary to the accepted belief of a transformation at the surface of the ground.

THE WINGED FORM.

The late rains of the summer of 1884, the season in which
nearly all the field studies of the pupa and winged form were
made, produced a generous supply of the white, club-shaped
rootlets, thus enabling us with properly arranged "traps,"
and bottles buried in the soil, to study the winged and other
forms. A beaker was also inverted over some of the bared
roots, and in one or two days an abundant supply of the
winged form was found flying about, and crawling upon the
side of the beaker which was most exposed to the light and
warmth. The young larvæ which left the roots for the
the smooth glass constituted a large proportion of the active
insects. A bottle which was in a cooler place, showed them
in far less numbers.

A steady and rapid production of winged individuals en-
sued from the beginning of the experiment, August twentieth,
through September, and a few were developed even later.
A large number of the confined winged insects soon laid eggs,
often as many as five for each individual. None of these
eggs, however, were observed to hatch; hence no sexual individuals were produced.

Migration of Winged Insects.—In arranging the glass jar
" traps " the soil was considerably loosened up, and thus was
prepared the way for the migration of the winged insects, which
occurred about the twentieth of August, when they could be
found in considerable numbers crawling about upon the small
lumps of earth, preparatory to taking wing. Only one was
actually seen to fly up to the vine, although others were
found quietly fixed upon the under side of some of the
leaves. This passing through the loosened earth, and later
through the unmolested soil, continued up to the tenth of

October, when the rains fallen a few days before put an end
to the development.

We had thus a continuous movement of this form coming
to the surface of the ground, not only from the loosened
earth, but as was seen later, also from the harder and unmolested soil. This was kept up until the fibrous roots were
destitute of pupæ, though still badly infested with young
larvæ.

After the discovery of the winged form in the University
vineyard, a considerable increase in numbers was noticed elsewhere. Never before had it appeared so plentifully as during the summer of 1884. While the peculiarly favorable
conditions of root growth found at Berkeley may not obtain
in all vineyards elsewhere, still equally favorable ones may
be presented. Yet, the more numerous cases of obvious
rapid devastation raise a strong presumption in favor of the
belief of a more widespread increase of this pest since, than
before 1884.

Movement of Young Larvæ through the Soil.—A peculiar
circumstance was noticed on the twenty-sixth of August, 1884,
in the appearance of a large number of larvæ upon the surface of the ground. They were found as much as two feet
from the stock, and from three to twelve inches from the
fine roots, as well as through the soil to the roots. The significance of their appearing in this manner can be appreciated
when we learn that they crawl upon bits of rubbish, sticks,
leaves, etc., upon the spot, and even take kindly to growing
canes placed in their way.

Just how far they can travel on the surface of the ground
in this manner we are not able to say, but certainly it increases the probability of their being transported upon boxes
and loose packages which are scattered so promiscuously
about the vineyard at that time of the year. It further shows
that the insect is not altogether dependent upon interlacing

roots beneath the ground for the means of spreading from
vine to vine. Small lumps of earth below the surface of the
ground, supplied with the smallest rootlets, were thoroughly
infested with the insect. Thus it is evident that the rapidity
of infection, or spreading, will surely be influenced by the
nature of the soil, *i. e.*, the greater or less facility with which
the insect can travel over it, or along the cracks in heavy
soil. In sandy soils the progress of the larvæ is very slow
and toilsome.

THE GALL LOUSE.

Up to August twenty-sixth, 1884, no specimens of the gall
louse, or leaf inhabiting form of the phylloxera, had been
identified at the University, or elsewhere in California, so far
as known. At that time the fresh young leaves near the
ends of three canes, which stretched from a "Canada" vine
towards an infested stock, bore a few peculiarly formed galls,
containing egg-laying mother lice as well as eggs, and numerous larvæ. A few isolated and abandoned ones were also
found on the old leaves nearer the stock of the vine. This
arrangement of a few isolated and odd galls nearest the
stump, and the peculiar fact that all the canes infested are
suckers coming from near the surface of the ground, suggests
the probability that the infection comes from the roots of the
vine rather than through other means. It is also noticeable
that one of these canes passes directly up through a portion
of the foliage, and still does not infect the adjoining canes.
Why the gall louse should appear just at this time, when the
conditions for the rapid production of other forms were favorable, and not at other times, is a question not easily answered. We are aware that similar freaks of change have
occurred in eastern experience in numerous localities, where
in 1870 the gall louse prevailed largely, the following year it
had almost entirely disappeared, or in some instances had
attempted, with more or less success, to locate upon other

varieties. The change during that same year even extended
to France, showing that atmospheric changes could not be
its sole cause.

Influence of Root Conditions.—It is more probably attrib-
utable to the influence of the root. During the whole investigation there has been noticed a very decided effect upon
the different forms, caused undoubtedly by the nature of the
roots upon which the insects are living. In our laboratory
experiments the larvæ are much smaller, more active or restless, and apparently more numerous than upon roots in their
normal state; our specimens being, of course, drier, and in a
poorer condition. The wing-pad insects, in the vineyard,
are formed only upon the smallest and most tuberous roots,
and in proportionately decreasing numbers as the roots become harder, scarcely ever appearing upon those which have
become tough and woody.

Identity of the Root and Leaf Louse.—Regarding the above
anomalous appearance of the gall-louse type, it should be
noticed that it has come upon a vine which has had no communication with any outside of those with which it has been
associated for years, and probably has had no way of becoming infected with any foreign type. If, as some maintain,
there be no direct relation between the two types, how is it
that the vine has borne them for a single year only, and that
they have not appeared again in 1885? They came at a
time, too, when we know that the temporary change of the
nature of the root system of the vine, caused by seasonal
peculiarities, had materially changed the nature of the other
forms produced upon them. It is said that climatic changes
influence, to a certain extent, the type which shall predom-
inate. If it holds in this case, it must be through the stimu-
lating influence of climate on the peculiar root growth which
made possible the development of a large number of winged
insects, which may possibly have been the means through

which the gall-type were developed; all extraordinary growth of vine having disappeared before the gall-type had been noticed to any extent. It at least seems probable, that the root-inhabiting form had changed its habit toward that of the gall-louse.

So soon as this opportunity of studying the relations of the two types were presented, an attempt was made to infect a clean cane of the " Canada " vine with the root-louse coming from the " Cornucopia " and appearing upon the surface of the ground.

A cane was bent from the opposite side of the resistant vine, and its terminal leaves fastened to an infested spot of soil. The leaves and part of the canes were soon covered with young larvæ, and a few quiet winged insects; the former passing freely about upon the leaves but forming no galls, or at least only doubtful or abortive attempts. Some of the young leaves upon the infested canes were pierced by young larvæ, which had settled just outside of the fresh galls, and had remained until a red dead spot had been formed. Others of the larvæ were seen crawling about; but they did not seem to establish galls. Contrary to the usual habit of the gall-louse, they kept mostly upon the under side of the leaf.

In fact, there were very few galls formed except upon the smallest leaves. One of these delicate leaves, an inch square, bore about thirty galls, a large proportion of which contained young larvæ, which were easily seen by looking through the leaf toward the light. Some galls even contained the mother, larvæ, and eggs.

It thus appears that, at least so late in the season, the change of habit from root to leaf is not readily made.

ENEMIES TO THE PHYLLOXERA.—Of the known *enemies to the phylloxera*, only two forms were identified during our observations. The *phylloxera thrips* were seen passing about in considerable numbers upon the leaves, and some even came from the galls, many of which they had cleared of their inhabitants. A few specimens of the *tyroglyphus*, or *phylloxera mite*, appeared among the winged insects that were taken from the "trap;" they were also found upon the roots of adjoining vines. It is, therefore, probable that its usual enemies have accompanied the phylloxera to California.

There have been other specimens of the same seen at different places, especially on roots taken for experimental purposes. Some were found on the fourteenth of April, the root specimen having been taken nearly a month before.

VARIETIES OF VINES BEARING THE SEVERAL FORMS.—In speaking of the forms found during the investigations, it must not be understood that all of the vines are productive of the same forms; nor must it be understood that when we speak of the rapid production of any particular form that this applies to all the vines infested. We are specially favored in this direction by having in our vineyard plot a few vines representing those varieties which are more or less resistant to either type of the phylloxera. Only one vine has developed the gall type, only one has produced the winged form in appreciable numbers, others only when transiently assuming the necessary conditions have produced them at all. If a slight growth of soft "tuberous" rootlets be formed we *may* find pupæ upon them; but upon the rootlets of the same vine having a firmer texture none will be found. Upon the "Cornucopia," however, this adaptability of rootlets to the production of the winged form extends to roots of a larger and firmer growth. The hybrid of the West's St. Peters and Clinton has produced all the forms we have thus far observed, except the gall type, thereby showing a special fitness for the production of the root-inhabiting types.

Here, too, we should note the peculiarity of these two vines, adjoining each other, hybrids of the same original

varieties, one producing the gall type and not specially
adapted to the root type, and the other producing all forms
of the root type with great ease and during the longest
period, but in no case bearing galls. Nearly all the observations in the past have been taken from this vine. It still
remains strong and vigorous.

DANGER OF SPREADING BY THE WINGED FORM.—The fact
that the winged form, so far as it appeared in considerable
numbers, was limited to a single vine in our vineyard plot,
necessarily diminishes greatly the probability of its spreading
to any other vineyard district from this place, and by this
form. If all the vines were equally productive of the winged
insect, the probabilities for infection would certainly be
greatly multiplied; but there would still be great doubts as to
whether they would be carried to any great distance and find
lodgment in a spot where the proper conditions for continued
life obtain. The winged insects have mostly been found
crawling upon the ground quite near the vine, being thus protected by the foliage above them from the winds which might
otherwise take them up into the air and transport them to
other districts. A few scattering ones only have been found
on the foliage, but the thick screen of trees and the ranges of
high hills in the direction of the regular winds prevailing at
their time of development, renders any actual danger from
this source exceedingly remote.

COMPARISON OF EASTERN AND CALIFORNIA TYPES.

As a basis for comparison of the forms which have come
under our notice with those known to exist elsewhere, I insert the following tabulated arrangement of the various forms
which this insect may assume, as presented in a report by
Professor Riley. It shows at once the complexity of its forms,
and the diversity of its habits:

I.—The gall-inhabiting type, forming galls on the leaves,
and presenting:

a. The ordinary egg with which the gall is crowded.

b. The ordinary larva.

c. The swollen, parthenogenetic mother, without tubercles.

II.—The root-inhabiting type, forming knots on the roots, and presenting:

aa. The ordinary egg, differing in nothing from *a*, except in its slightly larger average size.

bb. The ordinary larva, also differing in no respect from *b*.

d. The parthenogenetic, wingless mother, the analogue of *c*, but covered with tubercles.

e. The more oval form, destined to become winged.

f. The pupa, presenting two different appearances.

g. The winged, parthenogenctic female, also presenting two different appearances.

h. The sexual egg or sac deposited by *g*, being of two sizes, and giving birth to the true males and females.

i. The male.

j. The true female.

k. The solitary impregnated egg deposited by *j*.

bbb. The larva hatched from *k*, which, so far as is known, differs not from the ordinary larva, except in its greater prolificacy.

l. The hibernating larva, which differs only from *b* in being rougher and darker.

Forms observed in California.—In the course of our work we have met all the forms thus far known in the gall-inhabiting type; in the root-inhabiting type nearly all of the active forms represented in the table have been observed. They have been developed up to the production of what we have called the true sexual individuals, or, as designated in the table, the true male and female. The winged females which developed upon the roots and were caught in the trap, laid the sexual eggs, but none of them produced the individuals which would naturally have followed. No reason can be

given for the failure, more than the suggestion that the conditions may not be favorable in a glass vessel for the production of the form whose natural home is upon the leaves or
stock. There is, also, a possibility that these eggs may not
hatch even when under favorable conditions in this district;
for the winged insect has been repeatedly seen upon the stock
and leaves in a perfectly quiet condition, but no eggs have
ever been found with them. In fact, no eggs of any kind have
ever been found upon the upper part of the vine. When the
insects are confined in a vessel the eggs are soon laid.

Winter Egg.--The solitary egg, commonly called the winter egg has also not been found, although it has been the object of diligent search at all times. It is to this form that
much attention is being directed in European countries, as
its extinction offers a possible means of checking the ravages
of this pest. It is also supposed that a close relation exists
between this egg and the gall-type. If such a relation does
exist we should have found this form later in the summer upon the vines which bore the gall last year. A careful search
did not reveal it last winter.

*Probable Underground Development in place of the Winged
Form.*—The larvæ from the winter egg, of course, we have
not met as a product from the natural course of development
through the winged form, but their appearance must have
escaped our notice through the other parallel line of development which is accomplished entirely underground, and which
is described as differing only in its possible greater prolificacy
from the ordinary root louse.

This line of production may have been that from which
come the insects noted in a previous report (1882). Their
peculiar appearance at the time led to doubts as to what stage
of development they really represented. It was even conjec-
tured that they were the sexual individuals.

Professor Riley has omitted to mention in the table this supposed underground male and female spoken of by M. Balbiani, which does not pass through the winged state. According to this belief, a form similar to that produced by the winged insects may be developed on the roots and pass along the main trunk to the upper part of the vines and deposit eggs in positions similar to those selected by the winged form. This peculiar phase of development would assist in explaining some of the peculiarities regarding the continued prolificacy of the species in California where the true male and female appear so rarely, if at all. They may mingle with the common form which is so often found in considerable numbers two or more inches from the ground.

There is generally a marked distinction between the appearance of the young larvæ which are to develop either into the winged form, or are to become mother insects upon the roots. The latter being decidedly dull in habit, and pear-shaped, are quite easily distinguishable; the former are not so easily recognized until the wing-pads begin to appear but by this time they are already so far developed that they become less active, and in bodily form answer more nearly to the description of the wingless type. But then there has been noticed on some of our specimens among these larvæ, another form, which in activity and outward appearance closely resembles the undeveloped winged insect. It is of a very bright color, apparently smooth, and seems to separate from the remainder of the generation as soon as it is capable of moving. It is in fact never found in clusters, and mostly upon portions of the roots which do not show signs of having been attacked. These insects seem to be the explorers for the more sturdy productive ones which follow them. Their peculiarities place them undoubtedly upon the side of the winged form, and as there is presumed to exist a similar line of development, save the formation of wings, we may justly conjecture that this is the form corresponding in the biological series to the winged form, though never developing wings.

Hibernating Forms.—The hibernating larva is the final form of the year's development, upon which the future multiplication of the insect is dependent. This, and the winter eggs, serves to carry the insect through the winter. It seems, according to our observations, that this special precaution is not necessary, and is therefore not apparent in the California climate. Our winters are so mild that merely a cessation of work is noticeable, and not an extermination of the common forms. The young larva of the common egg-laying insect acts as hibernants without any apparent special preparation, and it seems to brave our winter without any trouble. The mother insect can also be found, although with difficulty. The larvæ themselves, in some cases, seem to have assumed some of the qualities of the true hibernants, for they are strong, and usually darker in color, often almost black.

In connection with the habits of the phylloxera during hibernation, an instructive sample of infested root was preserved this year, in an open vessel, filled with moist earth. The roo was protected from the earth and placed so that it could be watched through the sides of the vessel without being disturbed. Scarcely any change has been noticed since October fifteenth, when the specimen was prepared. Although the conditions have been quite similar to that of ordinary vine growth, except somewhat drier, there has been no movement to other parts of the root corresponding to the supposed movement of the phylloxera to lower roots when winter comes on. The insects have become somewhat darker, well developed young larvæ, with no appearance of the adult larvæ form. Up to February, 1886, there has been no appearance of reviving or moving about, still the extremities of the insect are extended and can be plainly seen as in their natural condition in summer. It is to be hoped that interesting facts may be gathered regarding moulting, and change of form, when they revive later in the spring.

In previous cases, where no soil was used and the temperature of the laboratory influenced them, eggs have been produced in mid-winter, and production continued until the root had decayed.

The soil of the above sample was moistened in December, and a good supply of grass roots and shoots started, but have since dried up.

It has never before been shown that the winter habits in California differed in the least from eastern countries. The winter of 1884–85 has shown that California climate is speially favorable to the life of the phylloxera. A period of three months will almost include the total time of inactivity, for we find active insects on the last of November, and newly-laid eggs on the first of March.

Moulting of Hibernants.—It is usually supposed that a certain number of moultings is necessary before the hibernants assume the mother state. This, I think, is only partially true here, and applies, if at all, to the youngest insects which pass as hibernants, and which were not fully developed when winter came on.

Numerous specimens have been carefully watched during the proper period, and no movement whatever was noticed until the insect began laying eggs. It was further observed that the abundance of young larvæ found at the close of the season in November were in about the same position, and apparently not changed in the spring when egg-laying began.

Sterile Winged Form.—The table speaks of two different appearances of the pupa and winged form. Undoubtedly this means the fertile and sterile kinds; the former alone has been produced during our experiments, although it was through the latter that the winged form was first found in California. These were observed in 1879, by Dr. Hyde of

Santa Rosa, and were identified by Professor Hilgard to be of the infertile variety. These are the only individuals of the sterile variety found thus far, and they came from rather large-sized woody roots, such as are usually found near the surface in ordinary California summers. May we not reasonably conjecture that the unusual summer rains of 1884, causing an unusual abundance of white surface roots, have also been instrumental in developing exceptionally the normal fertile winged form?

Mode of attack on Different Vines.—There is one point worthy of note as throwing some light upon the resisting power of vines; it is the manner of the insects' attack. In the common *vinifera* even, they show preference for particular spots on the roots, selecting those places where the bark is softest, usually near a crack. From this they extend upward and downward along the line where the tissue is continuous from that spot; and scarcely ever do we find them working at right angles from this line. When the sap begins to ooze out and rotting sets in, they precede it closely, always leaving a number of insects to continue the destruction until the spot becomes completely rotted and gives out no more sap. Large numbers of insects will often be found feeding upon such spots, apparently reluctant to leave them as long as any sustenance can be derived therefrom. So closely is this mode of working followed, that on many old Mission vines they will be found only on a single spot, while the remainder of the root is free from them. A root covered with a fuzzy bark is noticeably objectionable to them, a harder one with cracked or loosened bark is preferred.

Upon a thoroughly resistant stock the insects act quite differently. They are usually scattered about, apparently at a loss to know just where to begin operations. Their first piercings are made, and instead of a deep rotting which completely kills the bark to the woody tissue, a slight, thin blackening of the bark takes place, which does not extend further, and, if made on the finer rootlets, will often peel off, leaving the root perfectly smooth.

I abstain purposely from description of any chemical remedies, because I believe them too costly, and at the same time not effectual enough. They give us no guarantee, even if they could be so thoroughly applied as to exterminate all the insects, of permanent security; as they may at any time be again transmitted to the same vineyard, making continued applications necessary, generally with great danger to the vines. Only in cases when it is desirable to preserve a valuable piece of vineyard of a choice variety, it may be advisable to use Dr. Bauer's Mercurial remedy, which so far is the most promising, least dangerous, and cheapest of all that have been tried. Those who wish to try this, can find it fully described in Bulletins 18 and 48, which can be had from the State University on application.

CHAPTER VII.

RESISTANT VINES.

I have always been fully convinced, since I first studied the habits of the insect that in these we possess the only feasible and practical means of preventing and counteracting the ravages of the phylloxera. Insecticides, of whatever kind and description, are too costly in their application, and have to be renewed too often, to ever become practically applicable here or even in Europe. The lowest cost of their application of which I have seen an estimate, is about $30 per

acre, more than the general annual cost of cultivation, and
this is only a temporary remedy, which must be renewed
every few years, to be of use at all.   Besides, great care must
be exercised in their application, for an over-dose will kill or
fatally injure the vines.   The pest is liable to reappear at any
time, and thus it needs constant doctoring with costly remedies, to keep the patient even in a state between life and
death.   But when a vineyard is once established on American roots, of a variety suited to its locality, I believe it to be
fully efficacious, and European experience, as well as our
own *here* proves it to be so.   Any one who has seen the magnificent and flourishing vineyard of Messrs. Dresel and Gundlach, in Sonoma County, re-established on American roots,
when the vinifera had been totally destroyed on the same soil,
and the ground was full of the insect, cannot help to believe
them entirely and fully resistant.   During the five years that
I had charge of the Talcoa vineyards, near Napa, where the
insect is gradually destroying the old vineyard of 70 acres,
where I planted over 300 acres with American vines of different species, and replanted fifteen acres with American vines,
which had been destroyed by the phylloxera; I could fully
convince myself.   These vines, mostly wild Riparia, and Riparia
varieties, are now in their third summer, and although planted
on infested soil, and ground naturally not very rich, that had
been impoverished by over 20 years of constant bearing of
the vinifera, which occupied the ground before, but succumbed to the insect. The most striking illustration of the partial resistance of *all* American vines, was presented by some
old vines of Catawba, Isabella and Clinton, which had been
mixed in among the Mission vines and scattered among
them, about 50 in all, over perhaps two acres.   These remained fresh and vigorous, producing fair crops and good
growth every year, where Mission and other varieties were
utterly destroyed.

I do not wish to be understood, however, that certain species of American vines should succeed *everywhere* and in all soils. This would not be natural, and can hardly be expected. We have instances on record now, with even our short experience, which serve to show that we must again study the species adapted to our particular soil and locality. The Taylor and Elvira, two Riparia varities, succeed well at Dresels and Gundlach, below Sonoma, while at Kohler's Vineyard at Glen Ellen they do not succeed, while the Lenoir grows finely, and produces well; and the Herbemont again one of the most flourishing and vigorous at Gundlachs and Talcoa, does not succeed at Kohler's, only ten miles from there. So far, the wild Riparia, the Lenoir and Herbemont have given the most general satisfaction, while the Californica seems also to succeed well in all deep, rich soils, especially in adobe, and does not seem to flourish in dry soil or hard-pan. The same may be said in France and Europe, hence the conflicting reports from there about the results obtained with American vines. The question there is now, not so much their resistance, of which nearly all seem to be convinced, but the applicability and adaptability of certain varieties to certain soils and locations. The following, taken from the wine and spirit news, and published several years ago, may serve as an illustration of the estimation in which they are held there, and the extent to which they have been adopted to reinstate their failing vineyards. I have the pleasure to be one of the correspondents of Mdme la Duchess de Fitz James and gladly add my testimony to that of the gentleman referred to, as to her enthusiasm and zeal, and the incalculable benefits which her able writings and splendid example have conferred upon our beloved industry.

Under the heading " Measures for Combatting the Phylloxera," a pamphlet has recently been published at Bordeaux,

giving an account of a visit paid by M. A. Lalande, the
deputy for the Gironde, in company with M. M. Ed. Law-
ton and T. and P. Skawinski, to the districts of the Herault
and the Gard, for the purpose of studying the means employed in those departments with a view to the destruction of
the phylloxera, or where necessary reconstituting the vineyards
already destroyed. This journey, which extended over six
days, was undertaken more especially in the interests of the
vine-growers of the great and important district of the Medoc,
the centre of the richest vine districts of France.

Up to the present time the ravages of the phylloxera, although considerable, have not by any means been so serious
in the Medoc as in some of the other wine-producing districts
of France. In the department of the Gard, for instance, it
is stated that out of 255,000 acres of vines, 250,000 have
been destroyed ; while in the Herault, which produced at
one time 330,000,000 gallons of wine, and the average annual production of which was 220,000,000 gallons, the quantity for 1881 fell to 77,000,000 gallons only. From these
figures it will be readily seen that the field for inquiry offered
by these two departments was an extensive one, and the information to be obtained should be of extreme value as a
guide to other districts, and all the more so, as energetic
measures have been already adopted by the vine-growers of
the south, with a view, if not to save, at least to renew the
vines which constitute for them the chief wealth of their
districts.

Before proceeding further, we may say that the information and evidence obtained by M. Lalande and his fellow-
travellers throws a somewhat new light upon the question of
the phylloxera, and seems to show that, serious as the damage
caused by this insect has been, and still continues to be, the
case is not altogether a hopeless one. Of various remedies,
some thousands in number, suggested for combatting the

phylloxera, three only at the present time hold an important position. The credit of having suggested one of these, that of replanting by the American vines, is assigned to M. Laliman, and that gentleman shares with Baron Thenard, M. Dumas, and M. Fancon, the honor of having indicated to the French vine-growers the three means capable, according to situation and other circumstances, of resisting the terrible plague which at one time threatened to annihilate the vineyards of France — that is to say, the employment of insecticides, submersion, and American vines.

Of the first remedy we hear but few particulars during the journey undertaken by M. Lalande ; of the second no notice is taken at all ; while, on the other hand, of the results obtained by means of the third, most striking evidence is given, and, indeed, it is apparent that it is to the last remedy — that of replanting with American vines — that M. Lalande and his fellow-travellers attach the greatest importance.

On the first day of their excursion the chief interest seems to have been attracted to some vineyards in the neighborhood of Beziers, where an extent of more than 5000 acres of vines had been preserved for some years past by means of sulphate of carbon, accompanied each year by manuring over about a third of the extent of the lands in question. As a result of this treatment, it is stated that the vegetation was good and normal, although there were some points where the sulphate of carbon appeared not to have acted with the same efficiency and success as elsewhere. The failure in these cases, however, was attributed to the extreme humidity, which had paralyzed the action of the sulphate, a failure which, it is hoped, may be remedied in future by means of drainage.

Proceeding on the second day to the neighborhood of Montpelier, a visit was there paid to an estate on which all the French vines had been destroyed some time previously, and which now presented the interesting appearance of an entire reconstitution of the vineyard on a grand scale, by means of American vines planted ten years previously, and subsequently grafted with the French vines, which latter have since offered a perfect resistance to the attacks of the phylloxera.

In the same district another property was found where the vines, which were of a French species, had been grafted either on the Lenoir, Clinton, Taylor, or Riparia, and were in a splendid state of vegetation, with an abundant appearance of fruit.

Similar accounts are given as to a number of other properties visited on this and the succeeding days, as to which M. Lalande remarks : "It does not appear necessary to give a detailed account of all we have been able to observe. We limit ourselves, therefore, to remarking that, after having seen numerous specimens of all varieties of American vines introduced into France, we have especially noticed some Lenoir and Herbemont vines as presenting a magnificent appearance, with a fair quantity of fruit, although, it should be stated, much less so than was to be found where French vines had been grafted on American stocks." As a proof of this fact, some particulars are given of a property in the neighborhood of Montpelier where all the French vines had been destroyed by the phylloxera. Here some 200 acres had been replanted a few years previously with the American vines called Riparia, and these had been subsequently grafted with French vines. The results in this case were splendid, the vegetation being very fine, and the quantity of fruit enormous, in fact, all the vines were loaded with magnificent grapes, and these extremely well formed, so much so that the production had increased by half as much again per acre on the original yield.

Very much the same results were observed on the last day

of the journey, when visiting the extensive vineyards of the Duchess FitzJames. This lady has given much attention to the question of the advantages to be derived from replanting with American vines, and an article contributed by her some twelve months to the *Revue des deux Mondes* on the subject of American vines may be said to have contained, at that time, all the information to be obtained in regard to the same. Speaking of this property M. Lalande says : "We have much admired here the American vines — principally the Lenoir — cultivated with a view to direct production, as also the American vines grafted with French varieties. We have, however, still more admired, if this was possible, the energy and intelligence displayed by the Duchess FitzJames in the reconstitution of her vineyards. Some idea may be formed of this when we state that she has already successfully replanted 1,275 acres of vines. and is making arrangements for increasing this replanting to the extent of nearly 2,000 acres, thereby inspiring the conviction that the magnificent vineyards of this district — now almost entirely destroyed — will be able gradually to be reconstituted by means of American vines.

One other curious piece of information resulted from this visit. It appears that, it having been found that vines planted in sandy soils resisted better the attacks of phylloxera, these lands which formerly had been neglected, and were worth scarcely thirty-two shillings per acre, have now, after having been planted with vines, increased in value to nearly one hundred times that amount. Thus in the sandy soil of Aigues-Mortes the American vines which have been there planted presented a magnificent appearance with an abundant show of fruit.

From all the information obtained during their visit, M. Lalande and his fellow-travellers state, as the result of their experience, that they had found in the departments of the

Herault and the Gard the preference was given by the vine
growers almost exclusively to two kinds of American vines —
the Lenoir and the Riparia, although some other varieties,
such as the Clinton, the Solonis, the York-Madeira and the
Rupestris are considered excellent importations for grafting
with French vines.

Too much praise cannot be bestowed upon M. Lalande
and those associated with him in this journey, undertaken as
it was entirely in the interests of the French vine growers;
and if, as appears more than likely, as the result of their
visit, the practice should become general throughout France
of replanting with American vines as a means of resisting the
phylloxera, the destruction of the French vineyards, which
at one time appeared more than possible, may, it now seems
more than probable, be averted.— *Wine and Spirit News.*

This would seem to be conclusive testimony enough to
convince even the most skeptical, that vineyards can be rein-
stated with American vines.

I will now make a comparison of the different classes and
species, with their comparative value and applicability *here*,
as it appears to me after six years of close study and observation in this State. I shall take them in the order as they
seem to me the most valuable.

Vitis Estivalis—Summer Grape. I take this first, because
I consider it perhaps, the most important, at least some of
its varieties; and shall speak mainly of those which I think
most valuable, and which have already gained somewhat of a
reputation as direct producers, while they also make excellent stocks for grafting. I have already spoken of their leading characteristics in classification of grapes. They make very
strong roots, always at the lower joint, and are therefore eminently adapted to dry, and even stony hillsides, although they
also flourish in rich, deep land, but are averse to cold, wet,
hard-pan soil, as they do not like wet feet. But such soil is not

fit for grapes any way, and will never make first-class wine,
therefore should be avoided. They will stand the severest
droughts and always look fresh and green. Their thick and
persistent foliage affords excellent protection to the fruit, and
withstands, better than all others, the attacks of the white
thrip and other insects. They do not sunscald, and seldom
are attacked by mildew.

The *Herbemont* is perhaps the most valuable; a strong
grower, very productive and healthy. Synonyms, Herbemonts
Madeira, Warren, Warrenton. It was first cultivated by Mr.
Neal, a farmer in Warren Co., Georgia, who found the vine
in the woods, near his residence, as early as the year 1800.
It was afterwards cultivated by Mr. Nicholas Herbemont, a
Frenchman, at Columbia, and distributed under its present
name. He made wine of it for many years, which was
justly admired. Mr. Longworth introduced it at Cincinnati,
Ohio, and Mr. Charles G. Teubner at Hermann, Mo., 1847,
where I became familiar with it in 1852, and have cultivated
it in Missouri for twenty-five years. With winter protection
it seldom failed to produce a fine crop there, and I have seen it
produce 35 lbs. to the vine, or 2000 gallons to the acre. This
was with long pruning on trellis, and the crop was sold to the
wine makers at 7 cents per pound. It makes a very sprightly
white wine, if pressed lightly, immediately after being
crushed; and the remainder, if thrown into the fermenting
vat, and fermented from four to six days, will make a good
claret. The Herbemont has not been tried so extensively in
this State, on account of the prevailing rage for red wine, as
it lacks in color, but all the samples made from it have elicited
high praise for their sprightliness, easy and rapid clarification,
and their delicate flavor. The vine is a very strong, short-
jointed grower; wood gray, leaves deeply lobed, large and
thick, light green above, grayish below, which remain on the
vines, fresh and green, until December. Bunch compact,

shouldered, larger and heavier than any of its class; berry small, round, black with blue bloom, very juicy and sprightly, without pulp, well called by Downing, "bags of wine," skin thin. A very early and heavy bearer, has with long pruning produced 50 lbs. to the vine when four years old. An excellent stock for grafting, as it is an immense grower, and all the grafts of Vinifera varieties take well on it, but almost too good of itself to graft. Who ever plants it has the choice of keeping it for its own fruit, or grafting it. I should prefer the first. It propagates much more readily from cuttings grown in this State than in Missouri, and I have had from 50 to 75 per cent. to grow in nursery. The finest piece of Vinifera grafts Mr. Gundlach has are on Herbemont roots.

Perhaps the most valuable for direct production of the Æstivalis class is the *Rulander* or St. Genevieve. This is not the German grape of that name, but a true Æstivalis, first extensively cultivated by Dr. Koch, of Golconda, Ill., who obtained it from St. Genevieve, Mo.; vine a short jointed, stocky grower, wood short jointed, blueish brown; leaf heart shaped, not lobed; bunch small, shouldered, compact; berry small, round, brownish black, covered with blue bloom, juicy, sweet and spicy. Makes a very fine dark red wine here, resembling the finest Burgundy types, and is very productive. It is well adapted to spur pruning, on account of its short, stocky growth, and as it is easy of propagation, will be valuable for its own product, as well as a grafting stock. The must is very sweet and spicy, coming up to 28 to 30° Balling. Propagates easily from cuttings.

The *Louisiana* is very similar to the above, in growth of vine, leaf and fruit, but differs in making a lighter colored wine, resembling the delicate, high flavored and heavy hocks, It is as yet but little tried in this State, and only in my hands, I shall report fully on it in due time. So far, it is very promising.

Perhaps the best known of this class is the *Lenoir*. Synonyms, Black Spanish, Jack Grape, Jaquez, Devereaux. I sent it to this State, to Mr. H. W. Crable, of Oakville, Napa Co., in 1876, as he was desirous of obtaining a grape of very deep color, and abundance of tannin; and as I had tried it on a small scale in Missouri, I thought it would meet this want. It originated in South Carolina or Mississippi, and was first disseminated by a gentleman named Lenoir, hence the name, who grows it somewhat extensively in South Carolina, and made wine from it forty years ago, which was much admired. It is cultivated extensively in Texas, under the name of Black Spanish. It was introduced into France as early as 1864, where it is now perhaps more cultivated and its wine has a higher commercial value than that of any other American grape, on account of its intense color, and as its resistance to phylloxera has been fully demonstrated. It is a beautiful grower, with large, dark green, deeply lobed leaves; points of the young shoots reddish, wood brown, long jointed for its class. Bunch long, shouldered, loose; berry small, black, with blue bloom, juicy, no pulp, juice deep violet red. It has such a superabundance of color that one fifth of it will give the desired color to any light colored Zinfandel, and as it also has an abundance of tannin, it is likely to be a very valuable wine for blending, though I must say I do not admire it by itself as much here as I did in Missouri. It seems to me to be coarse and harsh. It is a very strong grower, and needs a six foot stake with abundance of spurs, to make it produce full crops, Specific gravity of must 28° Balling.

Cunningham. Synonyms, Long. This is perhaps the most valuable of the whole class as a *stock for grafting*, as it is an immense grower, takes the graft well, and propogates readily from cuttings. Though it bears very abundantly, the berries are small and dry, with a superabundance of acid, though also rich in sacharine. It originated with Mr. Jacob Cunningham,

Prince Edward Co., Va., about the year 1812. I received it from Virginia at Hermann, Mo., as early as 1852, and made some very fine wine of it in good seasons, though it seldom ripened its wood well, and proved too tender for that severe climate. Here it seems in its proper latitude, as it always ripens its wood. Vine, the strongest grower I know; wood, short jointed, blueish-red; bunch small, compact, shouldered; berry small, purplish-lilac, not very juicy, but high flavored and spicy. Leaf, heart shaped, not lobed, light green above, greenish white below, young shoots downy. I am inclined to believe it the most valuable stock we have for grafting, though not willing to recommend it as a direct producer.

These varieties all belong to the so called Southern division of the Æstivalis class. The Northern division, to which the Nortons Virginia and Cynthiana belong, have not, so far, proved a success here, though producing fine red wines in Missouri and Virginia. The climate seems to be too dry for them, the berries and bunches are very small and dry, and the leaves do not seem to be so healthy. I hardly think they will be worthy of cultivation here, especially as the wine they produce is not near so good as in Missouri and Virginia.

Perhaps the easiest and cheapest plan to grow the Æstivalis for *stocks* would be to sow their seeds. In connection with this I will say that the experiments I have made show a decided tendency in the case of the Lenoir to go back to the original type, the plants not showing as much vigor as could be expected from the parent stock. In the Herbemont seedlings, however, there seems to be a great tendency to sport, *i. e.* produce manifold varieties, but most of them showing great vigor and fine growth, while their root system was perfect; a large tap root, going straight down, with but few diverging roots, and these with the same downward tendency.

The seedlings made a very large growth the first season, and gave promise of being very fine grafting stock.

Vitis Riparia. Sand or River grape. Next to Æstivalis in importance as a *grafting stock*, I would place this, especially the wild Riparia, as it seems to succeed on a diversity of soils, propagates readily and easily, and takes the graft readily. Of the cultivated varieties there is but one, the Elvira, which is equal, if not superior, to the wild Riparia in this respect. The Taylor does not seem to take the graft well, nor is it as entirely resistant as the wild stock, as the wounds made by the insect on its roots, do not seem to heal over as rapidly as could be wished. The wild Riparia has the advantages of easy propagation, adaptability to location and soil, and uniting well with the graft. There is a difference, however, in the varieties; as the Green Hungarian, Marsanne, Franken Riesling, Clairette Blanche, Muscadelle de Bordelais, Pedra Ximenes and several others, take on it readily, while the Sultana, Crabbs Burgundy, Gamay, and Chanche Gris, do not unite so readily and surely. I have found very little difference in cuttings or seedlings, one year old seedlings will make about as strong a growth, if transplanted to the vineyard, as one year old vines from cuttings. If the latter are chosen, I would advise, however, before planting to cut out all the lower buds on the plant with a sharp knife, leaving only only those around the crown of the vine, or perhaps one below it. This will prevent suckering from below, will save much labor and disturbance of the graft, and one man can do the pruning of the vines for a set of four planters.

Vitis Californica. Our native wild species has been used to a greater extent as a grafting stock, perhaps, than any other, and it is certainly a very vigorous, strong grower, with a root system which seems to go down into the soil naturally. It has a very soft, fleshy root, however; fully as soft as the Vinifera,

and one which the insect, therefore, attacks readily. The
wounds seem to heal up, however, and the strong and vigorous habit of the vine may overcome its attacks. But it is
useless to attempt growing the Californica on very dry, stony
soils. The vine evidently does not feel at home there, nor on
hard pan alkaline soils; and as it makes but a very feeble growth
there, would readily succumb to the attacks of the insect.
Deep alluvial soils, and rich adobe lands seem to be suited
best to its wants. The wisest plan would be to choose the
Æstivalis for the former the hard pan alkaline soils, are
really not fit for any vine, and had better be left severely
alone for vineyard planting. As the subject is an all important one, I shall take the liberty of quoting largely, at the
close of this chapter, from the publications of Prof. E. W.
Hilgard, of the State University, who is one of the strongest
advocates of the Californica. As the vine does not propagate
as readily from cuttings as Riparia, the easiest and cheapest
way will be to raise seedlings, which make good plants for the
vineyard the same season.

Vitis Rupestris—Rock or sweet grape. As this is at home
on the most barren and rocky hillsides in Missouri, Arkansas,
and Texas ; it would seem natural to suppose that it was best
adapted to withstand our dry summers, and to succeed on the
driest soils. This seems not the case, however ; and its
growth, in such locations has not been superior if equal to
that of Æstivalis or Riparia, while in moist rather springy
soil, it makes a very heavy growth. It does not, however
take the graft as readily as these, suckers yet more, and I can
see no reason to recommend it when these can be had. If
planted it will be very important to cut out the lower buds
on the cutting or plants, as mentioned before for the Riparia.

Vitis Arizonica.—The wild vine of Arizona is closely related to this, and as I can see nothing in it to recommend, in

preference to the foregoing, and it does not seem to graft readily, we will pass it by.

Vitis Labrusca—Northern Fox Grape. This propagates readily from cuttings; but as it roots near the surface, and is not entirely resistant, although partially so, I cannot recommend it, except perhaps, a few of its varieties for immediate bearing and market grapes. For that purpose some of the earliest varieties may be valuable, such as Delaware, Early Victor, Wilder and Agawam. These could be grown on their roots, without grafting.

It can hardly be of practical value to enumerate the other classes, and I will only say in conclusion, that I can see no reason, with the phylloxera already in our midst, which would induce me to plant vinifera, when I can have resistant vines as cheap, as easy, and run no risk with them. If a years time is lost by grafting, the grafts grow so much more vigorously and produce so much more heavily than on their own roots, that the loss of time and expense in grafting will be more than made up in a few years.

The following extracts from the University report by Prof. Hilgard and others, will throw additional light upon this important topic. I only wish to add in the way of comment, that it did not take more than two seasons after planting the Riparia, either one year old plants from cuttings, or one year old seedlings, even in the unfavorable soil of Talcoa, to make them strong enough for grafting. Also that the early starting of the stock, referred to in the notes, did not seem to have any influence on the grafts, which did not start earlier than the same varieties on their own roots. But grafting seems to have a decided influence on coulure, as the grafts set and perfected their fruit much better than the same varieties on their own roots. Perhaps the partial obstruction to the flow of sap at the junction may account for this, as grafted trees generally bear earlier and more abundantly than seedlings.

The subject of vine stocks that will resist the attacks of the phylloxera, and can be safely used for the establishment of vineyards in infested districts by grafting, is one of growing importance to California, since upon these vines rest, at this time, the only hope of permanently maintaining vine culture in the most noted viticultural regions of the State. Although the progress of the pest is materially slower in California than in Europe, from causes adverted to in the last annual report, yet it is none the less certain. To ignore this fact is to imitate the fabled ostrich, hiding its head in the sand to escape its pursuers; and yet the indisposition to face the facts and prepare to meet the inevitable in the best possible state of defense is still so common, that the subject of the phylloxera is "taboo" in many places where he that runs may read the signs by the wayside, in the dying or fading vines that spot the vineyards; and he that calls attention to it is denounced as one who would "spoil the sale of the land." It is high time that this false and pernicious reticence and hiding-away should cease, and with it the useless expenditure annually incurred in the replanting of infested ground with non-resistant vines, or without any of the other precautions or preventive measures that would make such expenditure a reasonable business venture.

Since of late much has been said about resistant vines being after all non-resistants, upon the ground that in some cases they have succumbed, it is proper that the causes of of such occurrences should be placed in their true light, so far as the facts reported will justify conclusions. It is a matter of regret that it has not been feasible to undertake an exhaustive personal examination of such cases; but what is certainly known is sufficient to account for most of the well authenticated instances of failure.

The Meaning of " Resistant."— First of all it is necessary to dispel the illusion entertained by some, that resistant vines are such as are not attacked by the phylloxera. So far as our knowledge extends at this time, the insect will feed on any and all of the members of the true vine tribe (*vitis* proper) when occasion offers; but it is evident that some are better adapted to the taste or nature of the phylloxera than others, and are, therefore, more numerously infested when planted in the same ground with others; just as cattle will pasture on the sweet grasses in preference to the sour ones. The European vine (*vinifera*) appears, on the whole, to be the one most uniformly adapted to the insects' taste in all its varieties, and is always attacked in preference. It evidently offers the best conditions for the life and multiplication of the pest

It is not, then, a proof of non-resistance when a vine is found to be more or less infested; for, so far as we know, there are no true vines of which the phylloxera will not attack the roots when presented to them.

The true criterion is that the resistant vine and its roots will not only outlive the attacks, but flourish and bear remunerativecrops under the same conditions under which the more sensitive European vines will succumb.

But every vine, like every other plant, is subject to certain conditions of soil, climate, and atmosphere for its welfare. Any vine, or any other plant, may be planted where from unfavorable conditions it will not flourish, and where a slight addition to the adverse influences may cause it to either die or maintain only a feeble existence, useless to the cultivator for profit. The resistant vines are no exception to this general rule,

They have been planted, and expected to yield satisfactory results, where vines have been fruited for twenty or thirty years without the use of a particle of manure, and where, as a

result, the old vines, as well as the new "resistant" ones,
have died from sheer inanition.

They have been planted where no vine ever should be, if it
is to yield decent returns—in soils underlaid at a few feet by
impervious hardpan, and where the roots would remain
drenched in cold water until late in spring. They have not
resisted, as it was best they should not.

Cuttings, or rooted vines, have been planted in holes from
which dying phylloxerated vines had just been extracted.
They have found the cumulative pressure of having to take
root in fresh ground, and at the same time to feed a swarm
of half-starved phylloxera coming from the outlying roots of
the old vine, too much for them. They have failed to resist
what no young plant could be expected to survive under any
circumstances. Some have survived to the second, and even
third year, struggling against these adverse conditions, but have
finally succumbed, as they might have been expected to do
before. And again we hear of a damning example of the
failure of the resistant vines.

Adaptations of Vines to Soils.—But beyond such cases as
these, which are intelligible and avoidable under the guidance
of common sense alone, there is another class of reported
failures which is clearly referable to the want of special adaptation of the vine chosen as a resistant, to the particular soils
or location in which they were planted.

It is not reasonable to suppose that a vine which is naturally
at home in rich, heavy lowland soils, should not only flourish
but supply extra strength against attack from without, in thin,
meager uplands, or on land exhausted by long cultivation;
nor that a vine whose hardy roots resist the phylloxera when
growing in its natural location on dry, rocky uplands, will
necessarily retain this character when grown in rich, moist
lowland. To a certain extent, cultivation does modify and
equalize the natural soil-conditions, especially when it is thor-

ough and is faithfully kept up. But there always remains a certain margin of natural adaptation which must be respected even in the cultivated plant, and the more because climatic and seasonal conditions may render a strict fulfillment of the best culture impracticable or unavailing for the time being. Those cultivating adobe soils will appreciate the importance which this consideration may acquire, not only for one, but for several consecutive seasons.

Species and Varieties of Resistants.--Of the American species and varieties that, for practical purposes, may be considered as resistant under proper conditions of soil and moisture, the following are the most prominent:

1. *The Vitis Riparia*, or northern riverside grape and its cultivated varieties, of which the Taylor and the Clinton are the chief. The resistant power of the latter is now, however, pretty generally admitted to be inferior to that of their wild prototype, although they are better adapted to a great variety of soils. The *Riparia* is in its wild state emphatically a "riverside" grape, which in its natural condition ascends into the uplands only exceptionally, when these are unusually moist and fertile. Under cultivation, nevertheless, it does well on good upland, but is of slower growth than in its natural *habitat*. It does not frequent the heavier soils as much as the alluvial loams, of the upper Mississippi Valley. It is of very long-jointed, slender growth, so that its canes, while of great length and bearing abundance of foliage, are often borne by a surprisingly thin trunk, which is not as easy to graft as most other varieties. The cuttings root with great ease, but generally only a portion, varying according to the soil and seasons from one half to three fourths, are large enough to be successfully grafted the third year; seedlings arrive at about the same condition the fourth year from the seed. To offset these disadvantages the *Riparia* is now usually considered the

most generally and tenaciously resistant toward the attacks of
the phylloxera. It is very little liable to mildew.

2. *Vitis Cordifolia*, the southern riverside grape, so greatly
resembles the *Riparia* that for some time it was not distinguished as a separate species. While it is undoubtedly a very
resistant stock, the fact that it is at home in a region noted
for its perpetually moist atmosphere, seems to render it less
promising for general success in California than the *Riparia*,
over which, so far as known, it possesses no special advantages, save, perhaps in the case of very heavy adobe soils, to
which it is better adapted than the *Riparia*.

3. The *Vitis œstivalis* or summer grape is a native of the
uplands of the States east of the Mississippi, and is at home
on loam soils of good or fair fertility. It also descends into
the lowlands of the smaller streams, so that it and the *Riparia*
vine are not uncommonly seen side by side. But it is rarely
if ever found in the larger bottoms, though quite at home in
the lighter and usually well drained " second bottoms " or
" hammocks." Unlike the riverside grape, it objects to " wet
feet." It is a little subject to mildew. Of the cultivated
varieties of the *Æstivalis* grape, those of chief interest as
resistants are Norton's Virginia, Herbemont, and the well-
known Lenoir. The cuttings of these, as well as the wild
vine, root with some difficulty; they should be rooted in
nursery, and not in the vineyard itself.

A very striking example of the resistant powers of the wild
Æstivalis vine exists in this State, in the vineyard of John R.
Wolfskill, on Putah Creek, two miles from Winters, just within
Solano County. This case was alluded to in a previous re-
port (1882), but the stock was incorrectly stated to be Lenoir.
It has since been ascertained by Mr. W. G. Klee to be a wild
Æstivalis variety obtained at least ten years ago by Mr.
Wolfskill from Alabama, under the name of " coon grape."
It has a leaf much like the Lenoir, but bears a small, com-

pact bunch of sweet berries. Several hundred Muscat grafts were made upon this stalk when two years old, with scarcely any loss; and more lately some Huasco cuttings, obtained from the University, were similarly engrafted. Both are bearing heavily and regularly, while the *Vinifera* vines around have long since been destroyed by the phylloxera.

4. *Vitis Rupestris*, the sugar or rock grape of Missouri, is a very hardy vine, at home on rocky knolls and hillsides, where its wiry roots extract nourishment from the scanty soil and the crevices of rocks, in a climate already partaking somewhat of the aridity of the great plains. It would, therefore, seem to be of considerable promise for the foothills of California especially; of its resistance to the phylloxera there can be no question. It is, however, not easy to root from cuttings, being, in this respect, like the *æstivalis* varieties. In my personal experience I have found it to be of slow growth on rich upland adobe, even more so than the *Riparia*, so that when the top of the stock is sufficiently stout for grafting, that portion generally tapers off very rapidly downward, so as to afford very little "grip" for the graft, which has to be tied in very thoroughly. Whether from want of care in this respect or from the use of too many small stocks, my success in grafting the *Rupestris* the third year from the cutting has been very slight. The successful grafts, however, have shown a vigorous growth, and seem well joined. The multitude of wiry suckers which the stock persists in putting forth to the end of the season, constitutes an inconvenience, shared to some extent by the *Riparia*, and least of all by the *Californica*, which soon gives up sprouting its easily detached suckers. The *Rupestris* is least subject to mildew of all the resistant stocks.

Vitis Californica, the California wild grape (not, as some still imagine, the "Mission" vine, which is very sensitive toward the phylloxera), has been prominently brought forward

as a resistant stock for use in its native state, to the climate,
of which it must be presumed to be especially adapted. This
reasonable presumption gives it so great a claim to attention
and renders its preëminent success so probable, that nothing
but the strongest proof of its non-resistance should induce us
to relinquish its use. Even a cursory examination of its root-
habit shows that it understands the climate thoroughly. Two
or three strong cord-like roots start a few inches below the
surface of the ground, from a short but very stout trunk; and,
without branching or emitting rootlets, they go almost directly
down for from eight to twelve inches, according to the nature
of the soil. Then they begin to branch, but still with downward tendency, and without splitting into fine rootlets, until
they are fairly below the point to which the summer drought
is ordinarily expected to reach. Unlike the *Riparia*, its roots
are thick and fleshy, or cartilaginous, rather than wiry, and
one might suppose that it would invite rather than repel the
attacks of the tender-billed insect. The latter attacks it unhesitatingly, although it evidently prefers the non-resistant
Vinifera roots when these are within reach. The bites of the
phylloxera on the cartilaginous roots and rootlets of the *Californica*, however, do not result in the distortion which insures
the ultimate death and decay of the organs of the non-resistant vines so soon as they begin the process of turning into
wood (lignification). The wound will be found surrounded
by a raised ring which makes it resemble a miniature crater;
but the formation of this swelling does not materially deform
the soft root as it would a hard one. As it is well understood
that it is not so much the direct depletion caused by the insect's feeding, as the death of the roots caused by the distortion, that constitutes the fatal injury in the case of the *Vinifera*
stock. The cause of the resistance of the *Californica* is obvious enough. Here and there a rootlet, attacked by over-
whelming numbers, may be overcome and die; but if the vine

be placed under reasonably normal conditions of existence, it survives the loss so caused without any sensible effect either upon its general appearance or, what is most important, its productiveness. It goes without saying that the *Californica*, like any other vine, may be planted in the wrong place, where its half-starved roots become hardened, and instead of yielding so as to render deformation impossible, will curve and curl, and finally die and decay.

Among the many instances in which the *Californica* has satisfactorily shown its resistance to the phylloxera when planted on appropriate soils, may be mentioned that on the University grounds, where grafted seedlings, planted in 1882 in the holes from which badly infested stocks had just been taken, have ever since maintained a vigorous growth and abundant bearing. Also that of Mr. M. Thurber, in Pleasant's Valley, to whom, in 1882, some grafted *Californica* seedlings were sent from the University, for trial on his infested ground. They were planted among *Viniferas* dying from the attack of the phylloxera, that have since been removed. The vines on *Californica* roots, according to his statement made to Mr. W. G. Klee, are to-day vigorous and bearing heavily. With such facts before us, cases of reported failures require careful sifting before any conclusions are based upon them.

The *Californica* is very liable to mildew, and it is probably from this cause that it is but rarely found on the coastward slope of the coast ranges, which are much exposed to the sea-fogs. There is, however, no difficulty in protecting it by repeated sulphuring up to the time of grafting.

Vitis Arizonica, the wild grapevine of Arizona and Sonora, resembles somewhat the *Californica* in its general appearance and habits of growth. The leaves, however, are uniformly smaller and lighter colored, and more glossy ; the wood is of a light-gray tint, and the branches are very numerous and thin, with a tendency to the formation of long terminal runners.   In rapidity of growth it seems to be nearly or quite equal to the *Californica*, and quite its equal in resistance to the phylloxera, as well as to drought.   It is subject to mildew nearly as much as the former.   Its roots, also, seek the depths of the soil before branching, and the stock is stout and easily grafted.   From experience had at the University, it deserves more attention than has heretofore been bestowed upon it.

SPECIAL ADAPTATIONS OF THE SEVERAL RESISTANT STOCKS.

There is no reason why, in grafting grapevines, as great care should not be exercised in the selection of stocks adapted to the soil, and to the variety to be desired for bearing purposes as is done by orchardists here and elsewhere.

Just as every intelligent fruit grower will carefully consider when planting an orchard, what will be the stocks best adapted to his soil and locality, so the grape-grower must consider, so far as experience or other considerations can forecast it, which among the resistant grape stocks will be likely to do best in his vineyard.   An improper choice will be just as fatal to success in one case as in the other; there is no *one* stock that is adapted to *all* cases.   It is not a little singular that in this, as well as in some other points in the treatment of the vine and of its products, there should be a tendency to think of it as an exception to the general rules that govern in the treatment of other fruits; so that a sort of wholesale rule-of-thumb is applied to it that would be scouted in other cases by the same persons.   It has been claimed that not only the resistance of the Californian and other wild American stocks to the phylloxera has not been well proven, but that no perfect union between the *Vinifera* graft and the *Californica* stock is formed, and that the graft is liable to be blown over at any time; and finally, that if successfully grafted, there is no proof that such grafts will bear, or that the grapes will correspond to the quality of the scion.

As to the latter point, it may well be claimed as an established fact that the scion determines, in all cases, the character of the fruit, when any is formed. To deny this is to deny a fundamental axiom in horticulture, which has been demonstrated myriads of times for thousands of years. Minor differences may, it is true, arise from the habits of growth of the stock as compared with those of the graft when on its own root, whether as to rapidity of development, nourishment drawn from the soil, adaptation to climate, etc. In this respect the vine does not differ from other fruits, for which the best stock has to be ascertained by trial in each region.

As to the bearing of fruit, it is well known that, under certain circumstances of soil and climate, it may be greatly retarded, or even suppressed. But the grafts made on the University grounds on *Californica* stocks have all borne abundantly and early; and apart from many other examples of the same kind, Mr. Packard's experience in his one-hundred-acre grafted vineyard, *three years from the seed* (reported below), is a living example, than which a stronger cannot readily be found for other vines.

At the vineyard of Mr. W. G. Klee, in the Santa Cruz Mountains, near Alma, Mataro, Charbono, and Verdal grafted on the *Californica* stock three years ago, have been bearing abundantly ever since. Similar results have been obtained by many others.

As to the success of the grafts *when properly made*, the showing of 98 per cent. of successess in the case of Mr. Packard, cannot be easily excelled by grafts made on other stocks. As to the strength of the union, our experience here has been that when well made the junction becomes imperceptible and as strong as any other part of the vine.

It is true that when a strong grower is grafted upon a weak

one, there may be difficulty on account of the weak base of a
stouter trunk. But in the reverse case there is no trouble,
for a relatively stout base for a weak trunk is desirable. The
strong-growing *Californica* will, in its own home, furnish just
such a stock for all, or almost all, the *Vinifera* varieties, which
it exceeds in growth whenever planted in appropriate soils.

In my personal experience with the *Californica*, I have
found only one variety which seems to exceed it a little in
growth when grafted, viz.: the *Clairette Blanche*, which is an
extraordinary grower. In the case of five other varieties
grafted in my vineyard (Black Burgundy, Palomino, Mondeuse, Verdelho, Cinsaut) the graft junction is at the end of
the season either straight (*i. e.*, a cylinder), or like a wine
bottle right side up, the *Californica* stock forming the body of
the bottle, the graft the neck. There can be no doubt that
in these cases the stock will push the growth of the grafted
variety.

Where the same varieties have been grafted on the *Riparia*
or *Rupestris* stock, the case is just the reverse. Here the
graft junction resembles a bottle placed *neck down*, and it is
at least questionable that the stock will be able to supply fully
the needs of the graft, and pretty certain that it will not tend
to push the latter beyond what its growth would be if on its
own root. In the case of the *Rupestris* and *Clairette*, the
disproportion is painful to contemplate, the quill-like stock
appearing absurdly inadequate to the support of the graft that
has swelled to the proportion of a man's thumb. That this
is felt by the stock is apparent from the frequency with which
the grafts have thrown out their own roots when on either
Rupestris or *Riparia* stock, thus defeating the primary object
of grafting at all. No such tendency is seen in the same
varieties where they are on the *Californica* stock.

But this, it must be remembered, happens on a soil peculiarly well adapted to the *Californica*, and on which the *Rupestris*, at least, should not have been used at all.

Experience, which appears in some respects the reverse of my own, has been had by the Briggs Brothers, near Winters, as reported by Mr. W. G. Klee. The vineyard tract in question is located on Putah Creek alluvium and was planted with Muscat vines, which, notwithstanding the deep rich soil, began dying the third year from planting, being attacked by the phylloxera. A few years ago, a number of resistant vines were planted among the Muscats for trial as to their resistance and general success; they were, besides the *Rupestris*, *Riparia*, *Californica*, and *Arizonica*, a number of the cultivated American vines, such as Herbemont, Lenoir, Elvira, Taylor, etc. On the whole, all are doing well, but of the wild species the *Rupestris* is beyond question the strongest grower. Between *Riparia* and *Arizonica* it is hard to decide which is the most vigorous, but the *Californica*, while doing fairly well, is on the whole the poorest in growth, and on examination, the phylloxera was found to be apparently most numerous on the Californica roots. It was stated that in another part of the vineyard, not infested with phylloxera, the *Californica* is decidedly more vigorous than where it is among the infested vines.

It appears that in this case the *Rupestris* finds, in the more congenial lighter soil of the Putah alluvium, favorable conditions which are wanting in the rich adobe of Mission San José, yet without losing its resistant qualities, which on the whole, probably exceed those of the *Californica*. The latter represents among the vines what the plum stock does among stone fruits, while the *Rupestris* (and probably the *Æstivalis* varieties) are the parallels of the peach or almond stocks; and doubtless they are as little interchangeable as are the two orchard stocks.

It is obviously of the utmost importance that the conditions of the successful growth of the promising Californian vine should be fully understood by those contemplating its use as a resistant stock. Searching for a data on this subject, I was led, some years ago, to notice cases of particular luxuriance in its growth, and more especially those in which it ascends from its usual *habitat* in the moist lowlands, to the drier uplands. A very notable instance of this kind was fully investigated by me in 1884, and the results published in Bulletin No. 24, of the College of Agriculture, of which the relevant portion is given below:

No. 799. *Valley soil*, taken on a creek heading near Nun's Cañon, on the Oakville and Glen Ellen Road, Napa County.* The valley is a narrow one, of a briskly flowing stream on the Napa side of the divide. It is not under cultivation near the point where the sample was taken, but the spot is remarkable for the luxuriant growth of wild grapevines which cover not only the bottom, but run high up on the hillsides. The opportunity seemed a good one for ascertaining just what kind of soil the California wild vine delights in, thus giving a clue to the proper selection of soils on which it is to form the stocks. There is apparently little change in the soil for twelve or eighteen inches; it is of a gray tint, stiffish, and bakes very hard when dry, untilled; a light adobe or clay loam. For want of tools for digging, the soil was taken to the depth of eight inches only. It contains no coarse material, save a fragment of slate here and there. Its analysis resulted as follows:

VALLEY SOIL, SONOMA MOUNTAINS.

Insoluble matter	63.55	69.09
Soluble silica	5.54	
Potash		1.66
Soda		.22
Lime		.60
Magnesia		1.94
Br. oxide of magnanese		.11
Peroxide of iron		4.51

*On the occasion of a tour of observation, made under the auspices of the Viticultural Commission, October, 1884.

Alumina 13.71
Phosphoric acid... .17
Sulphuric acid... .07
Water and organic matter................................. 7.68
                                                          ────────
    Total 99.55
Humus 2.16
Available inorganic....................................... .49
Hygroscopic moisture...................................... 7.78
    Absorbed at...............................15° C.

The analysis shows good cause for the preference of the vine for th
soil, which is an unusually rich one in all the elements of plant food.
Its potash percentage is the highest thus far observed in California,
outside of alkali lands. Its supply of lime is not unusually large, but
still abundant; its phosphoric acid percentage is among the highest
thus far found in the State, as is, outside of marsh soils, that of humus.
In fact, any plant whatsoever might be well pleased with such a soil;
and the facts show that the native vine can be a rank feeder when opportunity is offered. These vines seemed to be young and had little
fruit set, but whether the latter point was an accident of the season, or
whether the soil is too rich for full bearing, requires farther observation to determine. If the latter be true, the remedy in such cases
would, perhaps, lie in the use of lime around the vines.

To the above conclusions should probably have been
added what seems abundantly obvious now, with a larger experience and scope of observation, viz : That the *Californica*, especially, seeks *calcareous* soils, which, on the whole,
are unusually prevalent in California ; and that its failure to
give satisfactory results in the well worn soils of Southern
France, regarding the calcareous or non-calcareous nature of
which we are without information, may readily be referable to
either of two causes, without prejudice to the resistant qualities of the native vine in the soils and climate of California.

So far as observations go, it appears that the *Californica*
is particularly adapted to *fertile* and *heavy* soils rich in *lime*.
In these its growth is certainly extraordinary, far outstripping
that of any other vine that has come within my observation ;
while in the equally heavy, but much less fertile soil of the
University vineyard plot at Berkeley, its growth has been

about the same as that of the *Vinifera* and other stocks on the same soil, .but uninfested by the phylloxera.

Rapidity of Development of the Several Resistant Vines.

That in our climate the *Californica* develops most rapidly of all, especially as to making a stock of grafting size, is hardly doubtful. The experiments made at the University from 1881 to the present time, as well as personal experience in my vineyard at Mission San José, fully corroborate the claim that the *Californica* is a stock of extraordinary vigor on favorable soils, and will bear very early grafting. It will be remembered that in the first experiments made with the grafting of seedlings at the University, in 1881, of seedlings one year old about forty per cent. were found stout enough for grafting, and were successfully grafted ; a thing not even remotely possible with any other species of vine yet tested, and least of all perhaps with the *Riparia*, whose seedlings are of exceedingly slow development. Thus, of a plantation of *Riparia* seedlings located on exceedingly favorable soil on Mr. John T. Doyle's place at Cupertino, not one could have been grafted when two years old, and only a few per cent. were fairly graftable when four years old.   .   .

At my own vineyard at Mission San José, the stocks from one-year-old *Californica* seedlings planted in spring of 1884, were without exception large enough to be grafted in spring 1885, despite a very unfavorable season. They were not actually grafted, however, until March and April, 1886, when, notwithstanding the extraordinarily dry season preceding, the trunks ranged in thickness from a minimum of two-thirds of an inch to fully one and a quarter inch, and sometimes more ; so that two grafts could readily have been inserted in a large portion of them. Of the *Riparia* cuttings planted at the same time as the *Californica* seedlings, few exceeded one-half inch in thickness, and very many were too slender to be grafted with any prospect of success, especially in view of

their large pith. Of the *Rupestris* cuttings planted at the
same time, few reached the thickness of half an inch, and
many appeared no thicker than when planted, three years before, except that they had at the top a short head, like that of
an old short-pruned stock, but too short to be of any use in
making the graft.

In this case, however, all the conditions were most favorable
to the *Californica* stock, as observed on the wild vine on its
own ground. The soil is a medium to very heavy adobe,
fairly to highly calcareous, and of a depth of several feet, underlaid by a gravelly "cement" and finally by a sandy substratum.
Analysis shows it to be rich in potash, phosphates, and humus,
and the experience of Mr. John Gallegos has shown it to be
specially adapted to the production of heavy-bodied and
deep-tinted red wines. The case is quite analogous to that
observed in the Sonoma Mountains, as quoted above.

A prominent case of remarkable "push" on the part of the
California wild stock is that reported by Mr. J. E. Packard of
Pomona, Los Angeles County, an account of which (published in Bulletin No. 45 of the College of Agriculture, October 9, 1885) is given below.

Bulletin No. 45.

REMARKABLE GROWTH OF VINES.

Scarcely more than two years ago Mr. Packard purchased, in different locations, two tracts of land of 170 and 86 acres, and immediately
began the improvement of the same. The tract of land consisting of
170 acres is situated four miles northwest of Pomona, on the San Bernardino road, and is of the very richest soil. Planting the main body
of the place to vines, the wild or native California grape was secured,
and this year grafted to Zinfandel, Burger, and Mataro varieties. The
growth made by the vines on this place is simply astonishing, as no
water whatever was used, and it is safe to assert that 98 per cent. of the
grafted vines are growing today, where, if cuttings had been planted,
scarcely one-half would have lived. In many cases by actual measurement, the canes are ten feet in length, and bunches of grapes weighing

three and a half pounds each have been picked from this vineyard.
From 80 acres about 25 tons of grapes will be realized this season, and
when it is taken into consideration that these vines have received no
water whatever, their condition proves conclusively that, in the right
soil, fruit can be produced without irrigation. This soil is no exception,
as there are many hundreds of acres of land in the Pomona Valley that
likewise need no irrigation whatever.—[*Pomona Progress*, August 20
1885.]

MR. PACKARD'S LETTER.

PROF. E. W. HILGARD, *Berkeley, Cal.* :

DEAR SIR : In response to your request I now send to you a copy of
the *Pomona Progress*, giving a description of the appearance of my
Californica vineyard. I will also make a brief memoranda of the details of my method of grafting them. I will here state that I grafted,
last spring, about seventy-five thousand, and have now a percentage of
loss of about two per cent. of that number.

First, the vines were cut off to within three or four inches of the ground,
and the brush hauled away; second, the land was plowed, the soil being thrown from the vines; third, grafting commenced February 10,
about three weeks before the vines started. For grafting I worked my
men in sets of about thirteen, as follows : One man to shovel dirt from
the vine; one man to saw vine at the surface, or one inch below the
surface of the ground ; three grafters—regular hands, who had never put
in a graft until they commenced this job ; one man following to wax the
union, who used a brush and wax pot; and, finally, seven men to shovel
the dirt to the vine, covering the graft to the top bud. All workmen excepting the grafters, were Chinamen. Each gang grafted eighteen hundred to
two thousand per day. Varieties grafted : Burger, Zinfandel, Mataro, and
Golden Chasselas. All have made a magnificent growth. Commencing
grafting February 18, I substantially finished three weeks after that date
—having something like ten thousand remaining, which were finished
up by two or three men by April 1, when the vines were in leaf. I can
see no material difference either in percentage of loss or in growth between the early and the late-grafted. The method used was a cleft
graft for the larger vines—say all larger than your little finger. For
the smaller ones a tongue graft was used, and a great many were grafted
which were not larger than a lead pencil. I find that the latter are
doing as well as any of the larger ones. As a matter of experiment,
one of my men cut the top of a vine off below a point where the roots
branched out, and inserted four Mataro grafts in as many small roots.

These four grafts are growing now, thus proving that it is unnecessary to graft in the crown.

I will mention the after-work when the grafting was finished. The field looked like a multitude of anthills at that time, on account of the dirt thrown up to the scion. I then plowed the land crosswise, throwing the dirt to the vine. Then, as soon as the union of scion and vine was strong enough, I cut the vine away, leaving one standard only, which I tied up. The "anthills" were leveled down, exposing the roots on the scion, which were cut off; and, as the union is at the surface, they cannot form a new, and the vine must be supported by the *Californica* root alone. The misses which I have I find to be almost invariably due to the fact that the scion was set with its sap veins entirely outside of that in the root, and as a matter of course such failed to grow. I used a great many lateral cuttings with an abundance of pith ; they all grew, however.

Of course, I used a great deal of care in keeping my scions in the best possible condition; they were never allowed to get into a position where they would dry out or injure in any other way. If there are any other points in regard to this matter which you desire to know, I will furnish them to you with pleasure.

JOHN E. PACKARD.

POMONA, October 2, 1885.

The above statement of Mr. Packard's experience with *Californica's* grafts is reproduced here, not as an example of what may ordinarily be expected, but of what may occur under extraordinarily favorable conditions. Its publication called forth at the time from Professor George Husmann, the well known writer on viticulture, the following communication, which was published in Bulletin 46 of the College of Agriculture :

TALCOA VINEYARDS, NAPA, October 24, 1885.

Professor E. W. HILGARD, *State University, Berkeley, California*:

DEAR SIR: As you desire reports about resistant vines, and grafting thereon, I will give a short *resume* of my experience here on perhaps the most difficult and varied piece of ground to be found in the State, being "spotted" with tough adobe, hardpan alkali, poor stony soil, and rich alluvial lands, and therefore a harder and more severe test for them than is ordinarily found.

The new vineyards at this place, comprising about 150 acres, were planted by me mostly in 1882. The varieties planted were for immediate bearing: Lenoir, Herbemont, Cynthiana, Rulander, and Norton's Virginia, all *æstivalis* varieties; and for grafting, about 10,000 wild riparia seedlings, 15,000 Clinton, and some few thousand each of other *riparia* varieties, such as Elvira, Missouri Riesling, Taylor, Uhland, Amber, Pearl, Marion, etc. Each variety runs in most instances from one end of the vineyard to the other, thus getting the benefit or disadvantage, as the case may be, of a variety of soils. In another piece of land we planted *rupestris* cuttings the same season, which also have about the same diversity of soils. I find a great difference in growth on the different soils, the most vigorous being on the alluvial and adobe, the poorest on the hardpan alkali. This may be considered applicable to all varieties planted, although the Herbemont seems to grow and succeed best on all soils. A piece of about an acre of the last named variety, planted in 1881, has been in partial bearing for two years, has always set its fruit well, and ripened evenly. The same was the case this year, when it bore a very heavy crop, many of the vines producing 40 to 50 pounds each, and ripening their fruit evenly and well, the must showing 24° on Balling's scale on the fifth day of this month. All the *æstivalis* varieties, however, need a 6-foot stake, and long pruning on canes or arms, to show their full bearing capacity. The same may be said of the Rulander or St. Genevieve, which set well and bore a splendid crop on three-year old vines, must showing 26° Balling the twenty-eighth of September, when we picked them. The Lenoir, Norton's, and Cynthiana set but a very light crop, owing, as I think, to the high winds which prevailed here all summer. They ripened early in September, Lenoir showing 27°, Norton's 30° Cynthiana 32° Balling, the must being of an exceedingly dark color, purplish black. All promise to make very fine wines, and as the vines are yet too young to show their full bearing capacity, I hope for a better yield next year.

In the spring of 1884 we grafted what was strong enough of the wild *riparia*, and the *riparia* varieties, although from the difficulties presented by the soil already mentioned, we had a very uneven stand. Our method was common cleft grafting, and has been described before We grafted on the wild *riparia* seedlings as follows: Sultana, Green Hungarian, Sauvignon Vert, Marsanne, and Franken Riesling (Sylvaner). A part of the last two varieties, five rows, were grafted on Elvira, running parallel through the blocks with the *riparia*. The great majority of the grafts took well, made a firm junction and a very strong growth where the vines were on favorable soil, but on the Elvira the success was rather more uniform and the growth stronger than on the wild *riparia*. The balance of our grafts were mostly on Clinton, which proved a much more satisfactory stock than I had anticipated, being in that respect as good as the wild *riparia*, and taking the graft readily. The grafts on Clinton were Herbemont, Lenoir, Pedro Ximenes, Chanché Gris, Traminer, Rulander or Gray Clevner, Petit Pinot, Gamay Teinturier, Mataro, and Grossblane. The Taylor although a very strong grower, does not seem to take the graft as readily as the three varieties named before, as our success was not as uniform and satisfactory.

The grafts produced some fruit last year, were pruned for bearing last winter, according to their strength, and most of them bore very heavily, with a great difference, however, in the same variety where they were exposed to the full force of the wind or sheltered by the hillside, the latter producing more than double. The yield was especially heavy on Green Hungarian, Marsanne, Sauvignon Vert, and Mataro. In a good many instances we gathered 30 to 40 pounds of Marsanne and Green Hungarian from a single graft, and the growth of wood for next year's crop is also strong and well ripened. The junction is so complete that it is hardly perceptible now, and the whole operation is a complete success. The bunches were very large and heavy, and the berries full size and of excellent quality, as a number of visitors, Messrs. E. H. Rixford and Wickson among them, can testify. Our wines, made from each variety separately, are fully fermented, and many of them clear now. When the time comes I shall take great pleasure in sending you samples of them for tasting and analysis.

I am fully satisfied that instead of losing time by planting resistant vines and grafting them, the grafts will bear more and earlier than the same varieties would do on their own roots, on account of the increased vigor caused by the stronger growth of the stock.

As to the alleged inferiority of the fruit and wine from such grafts, it seems too ridiculous for any one at all familiar with the laws governing horticulture, and the influence of the stock on the scion, to need refutation. In the case of grafts on vines, I have found, during a practice and observation of thirty-five years, that a stronger stock also imparts a more vigorous growth of wood ; and we all know that the more vigorous the tree or plant the larger and more perfect will be its fruit That such stronger growth also requires longer pruning to equalize the strength of the root and top is self-evident, but I have yet to learn that our growers would object to the increased yield resulting therefrom. In my opinion, the greatest perfection of the grape depends upon hav-

ing just as much to bear each season as it can ripen in perfection. If
we overload it, inferior, insipid fruit will be the result, and a feeble
growth of wood, which will also not ripen fully. If, on the contrary,
we prune too short, a rank, succulent growth, black knot, coulure, etc.,
will be the result, and the fruit will also suffer accordingly. On this
nice balancing of the powers of the vine more of the success of the
vintner depends than many are aware of.

That resistant vines planted on soil of ordinary fertility are and will
be a success I am confident beyond a doubt. That thousands of acres
have been planted to vines in this State which are entirely unfit for resistants or any other vines I am also convinced; and the sooner our
people learn that even a grapevine will not grow in waste and barren
places, too poor to produce even sagebrush, the better it will be for the
industry.

Yours sincerely,
GEORGE HUSMANN.

Loss or Gain of Time of Grafting.—While I am fully in
accord with Professor Husmann in respect to most of the
points made in the above communication, and believe that
the grafting on resistant stock should, and in fact *must* within a comparatively short time, become the rule instead of the
exception in California, yet I think his broad statement that
instead of *losing* time in bringing a vineyard into bearing
time will be *gained* by grafting requires material restriction.
I think the average experience will be found to be that there
is a loss of one year, or thereabouts, when a vineyard is
grafted instead of being allowed to bear directly from the
cuttings, and that on the large scale the cases of gain in time
will be very exceptional.

In the first place, it would be difficult to find a more vigorous and early-bearing stock than the Zinfandel grape, which,
were it resistant, could be recommended as a grafting stock
for its exceptional advantages in these respects. I doubt
that, in the case of this vine, even the *Californica*, grafted
successfully the first year from the seed, would distance it;
so that when this grape or others of similar habits are in ques-

tion, grafting on any other stock could be recommended only
as a matter of precaution against the phylloxera. But in the
case of varieties of weak growth, it may readily happen that
a genuine gain of time is secured by grafting on a vigorous
grower like the *Californica* or the kinds mentioned above by
Professor Husmann.

Proportion of Successes to Failures of Grafts,—Moreover,
it is altogether exceptional to find so large a percentage of
success in grafting as reported by Mr. Packard, above. A
loss of ten per cent. of the grafts made must ordinarily, I
think, be considered a very favorable result; it will more frequently amount to between twelve and twenty per cent,
varying not only according to the skill of the grafters, but
very largely depending upon the condition of the grafts used,
and upon the weather following the operation; also, to a not
inconsiderable degree, upon the nature of the soil. The influence of the latter becomes apparent from the fact that vine
grafting must be done, either several inches below the surface
of the ground, or, if done at or near the surface, the soil
must afterwards be piled up around the graft for protection
against drying out. In the case of an adobe soil, in which
the water may remain near the surface for several days after a
rain, the intrusion of muddy water into the cleft or cut, and
a consequent weakness of the junction and even risk of failure to unite, may take place. Hence, as such soils when in
good tilth retain moisture very strongly, I incline to think
that in them the grafting is best done within one or two
inches of the surface, the piling up of the earth around the
graft being relied on to prevent drying. With the same view,
my personal experience inclines me to favor late * rather
than very early grafting, because then the free flow of sap
from the stock keeps the graft in good heart.

* "Late" as regards the stage of development of the leaves, but not
necessarily late in the season.

A great deal, however, depends upon the condition of the grafts at the time the operation is performed. When fully dormant, they are of course, slower in coming, but less liable to injury from accidents of season than when somewhat advanced. In the latter case it may happen that moist weather following the grafting will push the buds too fast, before the stock has united sufficiently to fully support their growth, causing the buds to leaf out, and then, for want of proper support, die back to the main stem. From this condition the majority may recover, but a considerable percentage will fail to do so, or put forth but a weakly growth, leaving the grafter to lament a loss of twenty per cent. when, within a week after grafting, it appeared as though not one would fail to grow. If in this case the weather had been less favorable to rapid growth — that is, dry and cool rather than moist and warm — the loss would undoubtedly have been much smaller, as the growth of the scion would then have kept pace with the ability of the stock to supply the sap through a well formed callus.

While, then, a somewhat advanced condition of the scions — a swelling of their buds prior to grafting — may result very favorably when the grafting is done late, it involves a risk which is not incurred when they have been kept fully dormant.

Loss of Stocks from Graft Failures.—The grafting of a vine stock, as usually done, is a very severe operation for the plant. Were the graft not inserted so as to afford the stock ready-made buds for leaf-development, a great many of the weaker stocks would never be seen above ground again, as they mostly are through the formation of "adventitious buds," from which "suckers" sprout abundantly. When these suckers are persistently removed to the end of the growing season, very few stocks will retain life enough to sprout the next year. The majority will be killed by the exhaustion consequent upon the repeated effort to grow, unaided by the restorative action of the leaves.

While, therefore, the common practice of removing the first and even the second crop of suckers is a proper rule, in order to throw the sap into the scion as much as possible, yet so soon as it is definitely apparent or probable that the graft will fail, the "suckering" should be stopped, in order to insure a vigorous stock for regrafting the following season. From personal experience I am inclined to think that the necessity of close suckering, in order to make the graft "take," is commonly somewhat over-estimated; and that few grafts will fail altogether because the removal of the sprouts from the stock is omitted after the second time. From comparative observations on grafts treated differently in this regard, I incline to think that allowing the sprouts to grow will often so strengthen a dormant stock that it will push the scion into life, when, had the sprouts or suckers been removed, stock, scion and all would have perished.

But with the most careful treatment, and taking into due consideration the fact that a dead stock involves for its replacement a loss of two or three seasons, while a stock whose powers have been judiciously husbanded may be successfully regrafted the following year, yet a certain percentage of loss will thus take place, involving the replanting of a cutting or seedling. This, with the graft failures, defers the completion of a full "stand," and counts in the matter of delay in bringing a grafted vineyard into full bearing.

Considering the advantages to be gained by grafting on vigorous resistant stocks in regions menaced by the possible importation of the phylloxera, one season's delay (which I think should be looked for by those who graft) should not deter any one from taking this needful, and with proper care as to adaptation, ultimately profitable precaution.

Crop from Grafts of the Same Season.—When bearing

wood is used for scions grafted upon vigorous stocks, a crop
will generally be borne the same season, sufficient to give
some insight into the adaptation of the grape varieties to the
local climate. The crop will, of course, be belated. If it
matures well, nevertheless, it is positive proof that it will do
so 'still better when older; the berries will have less sugar and
more acid than in succeeding years. If, nevertheless, these
amounts are fairly satisfactory, it will convey the assurance
that when older the vines will yield a good product. It need
hardly be added that the older the stock the more nearly the
results so obtained approach the average result of future years;
yet excessive bearing of grafts should not be permitted, in order to avoid weakening so as to spoil the "good half crop"
that may be expected the second year.

Propagation of the Resistant Stocks.—Experiments on this
important subject were made at the University some years
ago, and the results were published in October, 1885, in
Bulletin No. 34 of the College of Agriculture. The stocks
experimented on were the wild species *riparia, æstivalis, Californica*, and *Arizonica*, being those deemed of the greatest
importance at the time. The following is the report of Mr.
W. G. Klee, then head gardener of the agricultural grounds,
on this branch of work:

In a previous report (1882) an account was given of some experiments in growing wild grapevines from seed, as also of the influence of
carbon bisulphide upon the cuttings disinfected by means of its vapor.
It was deemed desirable to institute comparative experiments on the
facility with which the native Californian and Arizonian vines can be
grown from cuttings. For the sake of farther comparison, cuttings of
the summer grape (*Vitis æstivalis*) and of the Northern and Southern
Riverside grapes (*Vitis riparia and cordifolia* respectively) were also
planted under precisely similar circumstances. All the cuttings were
taken from vines growing in the garden of economic plants; and without exception, wholly of the previous season's wood, which was very
thin, being on the average not more than one-sixth inch in diameter.

The cuttings were all made about the same time, viz: at the end of

December, 1883 ; and were cut eight to nine inches long. Soon after they were put under ground in a shady place, and there remained until planting time.

The soil of the nursery bed being rather heavy loam, its condition was improved by a heavy dressing of barnyard manure, and with the subsequent addition of fifty pounds of quicklime to an area of 20x30 feet, or about 1.72 part of an acre.

At the beginning of April the bed was deeply dug with forks, and on April 6 the cuttings were planted, some sand being spread in the bottom of the trench to facilitate rooting. Drills were placed 18 inches apart, and cuttings 4 to 6 inches in the row, two eyes being left above the ground, the lower one just at the surface.

The *Vitis riparia* was the first to start, and was followed, in eight or ten days, by the *Arizonica;* the *Californica* being the last, fully three weeks after the *riparia*, and starting quite slowly. The *æstivalis* started a trifle sooner than the last mentioned.

During the season (which, as will be remembered, was quite a moist one) the bed did not receive any watering, but was kept clean and well worked.

Small as these cuttings were, their growth has been very good, and as shown in the table below, a large percentage of all the varieties rooted ; each kind exhibiting its peculiar habit of growth.

The *riparia*, which started first, was also the first to stop, the leaves all turning yellow at the end of September. The *Arizonica*, at the same time, showed signs of having made all its growth, but kept a good green color; while the *Californica* still continued to grow vigorously. The same order, precisely, was observed in the 4-year old vines from which the cuttings had been taken, so that it doubtless represents fairly the respective habits in this climate.

The bed at the time presented an interesting sight ; the *riparia* with its long spreading canes and fading color contrasting strikingly with the bush-like, upright habit of the Arizonian vine, and both with the running but more robust habit of the Californian. The leaves of the latter only yielded to frost and remained on the canes until spring. The *Arizonica* dropped them soon after the first frost.

The following table shows at a glance the main points in the growth of the several species:

108GRAPE CULTURE AND

Name.	Per Cent of Cuttings Growing.	Average Length of Canes— Feet.	Diameter at Root Crown— Inch.
Vitis riparia	95	3.1-2	3-8ths.
Vitis Arizonica	97	2.0	5-16ths.
Vitis Californica	85	3.0	7-16ths.
Vitis æstivalis	85	2.0	3 to 4-16ths.

The roots of the cuttings exhibit the same striking differences observed in the seedlings of the same species. The Californian vine has by far the most vigorous roots, as well as the smallest number, and these strike directly downwards. The roots of the *Arizonica* are next in vigor, and also strike straight downwards, but are much tougher in texture. The *riparia* has a great profusion of roots, but of a much more spreading habit, apparently seeking to remain near the surface—a habit it always retains when older, and to which its early starting may in part be due. The *æstivalis*, although apparently the smallest and weakest grower, yet developes a powerful root system, with a more downward tendency than the *riparia*, and in deep soils, or where roots can penetrate deeply into the substratum, it should do well. Its roots are, during the first few years, stronger in proportion to the top than is the case with any of the other species tested, and this speaks strongly in its favor for use as a grafting stock.

As it is in many cases of importance to be able to distinguish the canes of the several wild species from each other, I call attention to the distinctive characters afforded by the configuration of the pith at the nodes or eyes, when a joint is cut lengthwise; a subject to which attention was first called by the late Dr. Engelmann, and of which examples referring to Eastern species are figured in the Bushberg catalogue for 1884. In these figures the pith of the *riparia* shows at the node a thin cross partition; in the *cordifolia* or Southern riverside grape, this partition is quite thick. In the Southern Muscadine or Scuppernong grape, (*V. vulpina*) the partition is entirely absent; while present, in varied forms, in all other American species, The difference between the *V. Californica* and *Arizonica* in respect to the partition is almost precisely the same as that between the *cordifolia* and *riparia*, and will serve to distinguish the cuttings from each other; the cross partition in the *Arizonica* being quite thin. Figures are, of course, needed to illustrate these points more exactly, but when once noted they are easily recognized.

W. G. KLEE.

The above record of observations made by Mr. Klee are confirmed by the experience of others, had during the past season. As regards, first, the rooting of *Californica* cuttings, the same percentage result as to success is reported by Messrs. Coates & Tool of Napa, who state that while they have had little success with cuttings from wild vines, they are well satisfied with the outcome from cuttings made from cultivated *Californica* stocks. Mr. J. H. Wheeler reports similar success. The same seems to be true of the *Arizonica*, which was at first reported to root with great difficulty. It is true that the season of 1884 was an unusually favorable one for the rooting of cuttings; but in the nursery the same conditions can be artificially kept up at any time; and in 1885 the results were as good as in 1884.

Of the above varieties the *riparia* is the one of which the cuttings can unhesitatingly be planted directly in the vineyard; the *Arizonica* is next; while the rest, including the *rupestris*, should preferably be rooted in nursery. It is true that the percentage of successful rooting of *Arizonica* is in the above table practically the same as that of the *riparia*; but the extreme thinness of the cuttings of the former renders them so much more delicate in handling that only experienced hands could be trusted with planting them in the vineyard, and from the same cause, their vitality is easily injured by exposure to drying-out, such as is but too apt to occur in the wholesale method of planting.

From what I have seen of the growth and habits of *riparia* seedlings, I should not incline to their use in preference to cuttings, on account of their delicacy and extremely slow development. It is quite otherwise with the *Californica*, of which even very small seedlings grow very readily and rapidly, and are very tenacious of life under very adverse circumstances. In regard to the latter point, I state that in my own planting, a bundle of about two hundred moderately sized seedlings were, by the carelessness of a workman left exposed in an open field, with only a doubled jute sack to cover them from the sunshine, which prevailed quite hotly during at least half of ten days during which they remained in this position.

When discovered, some of the smaller ones on the outside of the bundle were completely dried, but the majority were only somewhat wilted, and fully 80 per cent of them lived after planting in the nursery. This tenacity of life is a strong recommendation for the seedling *Californica*, as most likely to insure a full stand, even under conditions that would seriously diminish the percentage of success with even the most easily rooted cuttings.

Differences in the Earliness of the Several Stocks.—An important point of which the exact influence cannot yet be foreseen, but which deserves serious consideration, is the relative earliness of the several resistant stocks. However little the stock may *specifically* influence the character of the fruit, it is probable that one and the same grape variety grafted on the *riparia* on the one hand and on the *Californica* on the other, would be materially influenced in the earliness of its start in spring, as well as in the maturity of its fruit, by the roots upon which they are severally dependent for the rise of the sap. The *rupestris* is even a more extreme case than the *riparia*, for this spring it has started on an average at least one week in advance of the *riparia* on the same soil, making possible, according to Mr. Klee's estimate, a difference of nearly four weeks as between *Californica* and *rupestris*. In my vineyard at Mission San José, the actual difference this season has been about three weeks.

This consideration becomes very serious in relation to damage from frost, which would be likely to be much greater on *riparia* roots than on those of the *Californica*. Again, as regards the ripening of grape varieties which it would be desirable to blend, and which yet ripen too far apart in time to be fermented together, it might be practicable to retard the one and advance the other by judicious selection of the stock, so that both should ripen nearly or quite at the same time. Similarly it might be feasible to make the same grape variety come in at two or more different times, so as to diminish the rush of its vintage, and enable us to use it for grape blends, in different combinations which otherwise would not be possible until after fermentation. Excellent opportunity for observations on these points will be presented at Mission San José during the season of 1887. Thus far the present season's experience, and that of others who have had opportunity for comparative observation, does not tend to show as great a difference as might have been looked for.

CHAPTER VIII.

GRAFTING THE VINE.

I hardly need call attention, after the foregoing chapter, to the importance of this operation, on which so much of the success of our vineyards depends. But it is not alone the advantage it gives us in transforming a non-resistant vine into a resistant one. Its advantages are manifold, and may be summed up as follows :

1. The facility by which new and rare varieties may be rapidly increased, by grafting on old, healthy vines, often making a growth of from 10 to 20 feet the first season.

2. The short time it takes to fruit new and untried varieties, as the grafts, if strong enough, will bear a few bunches the first season, and nearly a full crop the second.

3. The facility by which vines of worthless or inferior varieties can be changed into valuable bearing vines of superior fruit.

4. Varieties which will not grow readily from cuttings will generally graft easily, and can thus be propagated faster.

5. Most important of all, it gives us the means to successfully combat the phylloxera, as we can change a non-resistant vine into a resistant one, by grafting on a resistant root near or above the surface; or by reversing the case, grafting eight or ten inches below the ground resistant cuttings on non-resistant roots, when the scion will take root at the junction, thus transforming itself into a resistant vine in time; while the stock will furnish the sustenance temporarily, to make it grow rapidly and vigorously.

All these advantages are so great that they will be obvious to any one. Being convinced of its importance, we will now consider *when* and *how* the operation should be performed. I shall not try to worry and confuse my readers with many complicated methods, but only mention a few of the simplest. Although the vine may not graft with the same ease as some other fruits, as the cambium and inner bark of the vine is very thin, yet it presents no serious difficulties, and if properly performed, from 75 to 90 per cent. of the grafts will grow.

First, as to the proper *time*. Although it may be done in this State as early as February; yet, if the scions are left dormant, I would rather wait until April or even May, although this will vary with the seasons and location. If the sap is flowing rapidly at the time, no matter ; the junction will be formed all the more readily, provided it is done rapidly and .well, so as to avoid exposure to the air in stock or scion. A lot of grafts were put in by me the first week in April this season. As I was called away by business, I could not finish until ten days later, and a few vines of new varieties were grafted still ten days later than these. Strange to relate, the last started first, the second lot next, and the first last of all. At the present date, June 28, many of the grafts have made a growth of seven feet, and show from three to nine bunches of fruit. These are Marsanne, grafted on four-year-old Charbono, Vinifera on Vinifera. This may serve as an illustration. About 90 per cent. of the second and third lot are growing; of the first, about 75 per cent., and only now starting into vigorous growth. All the scions were kept dormant, by being buried in a shady place.

Next, as to the proper *choice of the scions*. This I consider very important. The scion should be of medium sized, short jointed, firm wood, with well developed buds, and, of course, well ripened. The large canes are inconvenient, and generally too loose and pithy in their texture to make good scions, while the small wood has generally only a single bud, which is easily rubbed off and therefore liable to fail. About the size of a common lead pencil will be best, though somewhat larger scions may be used on heavy vines, and smaller ones on correspondingly small vines. Here the good sense of the grafter is the best guide, and a little practice will soon make perfect.

As to the *best* methods, they will *all* succeed; if they fulfill one great requisite, perfect union of the inner bark in stock and scion. As this is thicker on large stocks than on small ones, and comparatively thicker on stock than scion, it becomes self evident that the scion should be set deeper below the rough outer bark of the large stock, than the comparatively thin bark of a small one. With these few general hints, I will now describe a few of the simplest, and most common as well, as most successful methods.

COMMON CLEFT GRAFTING.

This is done by cutting off the stock horizontally, at some smooth place below the crown or the ground. I prefer to have about an inch of smooth wood, above a node or joint in the vine. The reason for this is, that the cleft of the stock ought to be about an inch long, and in splitting, the intervening node, (or whorl of roots) will prevent it from splitting farther, as it will then close well around the graft. Now split the stock longitudinally, with a sharp knife, chisel, or pruning shears. In grafting stocks not larger than an inch in diameter, I use the shears for both the horizontal and longitudinal cut, as on these, I insert but one scion; I choose the smoothest side for its insertion, keeping the blade of the shears on the side where the scion is to be inserted. (Fig. 6.)

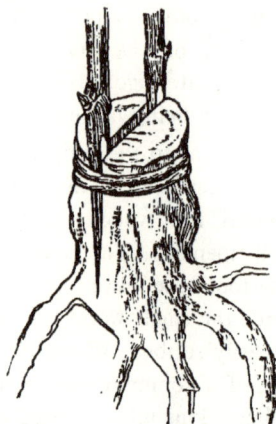

FIG. 6.

This will prevent bruising of the bark. Then I prepare the scion. With a sharp knife, I cut a simple wedge (Fig. 7.) beginning at an eye or beed, and cutting a long sloping cut toward the middle, and a similar one on the opposite side. The side of the wedge should be thickest at the eye, and thinner toward the other side. Now open the cleft where you wish to insert the scion, and push it down firmly until the bud is even with, or just above, the upper surface of the stock, taking care to fit the inner bark of the scion closely to that of the stock. If the stock is large it may require a small iron wedge to open it, bent to one side, (Fig. 8.) and to insert two scions, one on each side. If the operation is

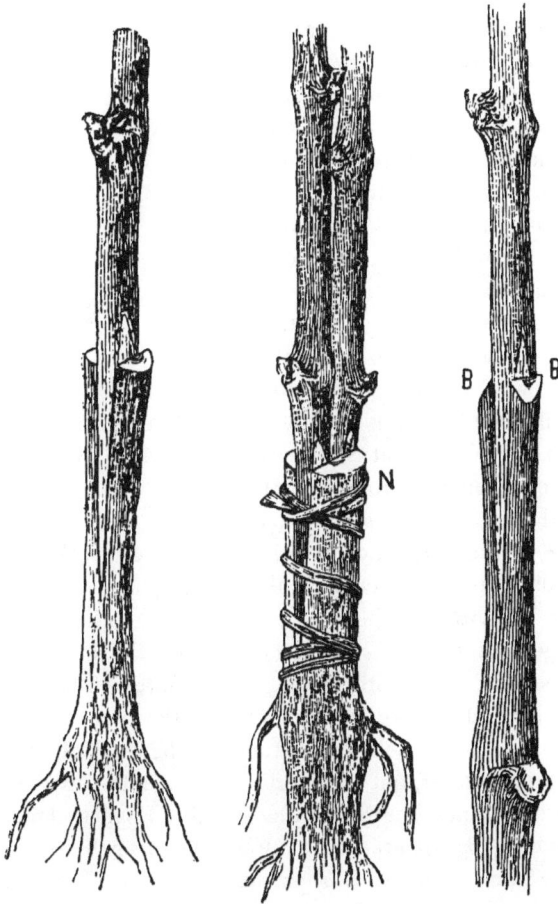

FIG. 7.

performed below the ground, as is generally the case, and
the stock is strong enough to
hold the scion firmly, no bandage
is necessary. A little moist earth,
pressed upon the cut of the stock
and around the wound, is all
the sealing it needs. But if the
stock is small, it ought to be firmly
tied with raphia, or strands of rice
straw, as found in the matting
around tea, which makes a very
good bandage. Draw the cut
firmly together, and wind the
wrapping around the stock evenly
until the whole cleft is covered.   FIG. 8.
I generally take three buds to the scion.   If above ground, it
ought to be waxed, that is the whole cut covered with grafting wax of some kind, to exclude the air.

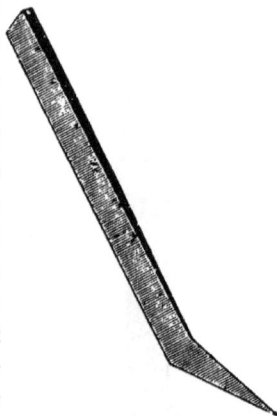

A variation of this method is. to make a slanting cut in
the side of the stock, downwards to the middle, then cut
your scion to a simple wedge as above, and push it down on
one side, so as to join the bark of the stock.   This has the
advantage that the stock need not be cut off, in case the
scion does not unite with it,. and as the fibres of both, the
scion and the stock, are cut transversely, the pores join better.   As soon as the scion starts, cut off the stock above it,
taking care, however, not to disturb the scion in the operation.

Another common method is the so called English cleft, or
splice graft, (Fig. 9) especially applicable to smaller stocks,
when the stock is not much larger than the scion. A sloping,
transverse cut is made on a smooth place of the stock, upwards, and a similar one on the scion downwards, then a
split is made longitudinally, from the middle of the transverse
cut ; in the stock downwards, upwards on the scion ; and

the upper wedge of the scion, thus formed, is pushed into the slit on the stock, until both fit close-ly, and the lower end or lap of the scion rests closely upon the lower bark of the stock. Here also, care must be taken, that the inner bark of stock and scion fit well. It is then bandaged as the cleft graft.

A modification of this is the so called Champin graft, called so from its inven-tor, Aimè Champin, but I cannot see that it is superior in any way to common splice grafting.

Fig. 9.

There are a great many other methods, as saddle grafting, grafting by approach, inarching, etc., but I wish to confine myself to only the simplest and most generally successful.

This is the "modus operandi;" now for its practical appli-cation in the vineyard.

In grafting in this State, we generally have one or the other of the following objects in view, viz.:

1. Changing worthless vines into something more valuable.

2. Grafting noble vinifera vines on resistant stocks, to in-sure them against phylloxera.

3. Grafting the most valuable resistant vines, for immedi-ate bearing, on non-resistant stocks.

We will commence with No. 1 and suppose a case of a four-year-old vineyard of Mission or Malvasia, to be con-verted into Semillion or Petit Syrah. Choose your scions as indicated before, then graft them under ground, at the first smooth place you can find. I would cleft graft these, or use the modification of side grafting, as mentioned before. Di-vide your labor as follows: Let one man dig away the earth from the vine, until he comes to a smooth place, then saw or cut off the vine about an inch above a node or whorl, also making the longitudinal cut. The grafter follows, cutting the scions and inserting them. He can carry his tools in a basket, for greater convenience, keeping the scions in a moist cloth, and ought to be the most careful one of the lot, for on him depends the success mostly. A third follows, pressing moist earth on the top of the stock and around the cut, and then filling up with finely pulverized earth to the top of the scions. If large enough to hold the scions firmly, the stocks need not be tied; if small, they should be; if very large, two scions ought to be inserted, one on each side of the stock. I need hardly mention that care should be taken not to move the scion in tying.

We come now to the second case in point, grafting a young resistant vineyard with non-resistant vines. In this case, we will suppose the vines two or three years old, from half an inch to an inch in diameter. These we can graft above or below the surface, as we may choose. Grafting below the surface no doubt will be more successful, as there is not so much danger of drying out. Still this can be overcome in a measure, by making a false surface ; that is, by making a mound of earth above the ground, and around the graft, and there is certainly less danger in removing suckers to disturb the graft, so that, on the whole, the advantages may be evenly balanced. In any case, we select a smooth place for insertion, as before described, and either cleft or splice graft; for larger stocks, I prefer the first, for smaller stocks the latter. If grafting below ground, and a force of six men is at command, they can be utilized as follows: One to dig around the vine, to take away the soil to a smooth place ; another to follow with a sharp pair of pruning shears to cut the stocks horizontally and longitudinally, to receive the graft. If the cut is made somewhat transversely across the fibers, so much the better, as it will increase the chances of a union. This can also be modified by leaving the stock,

making only the transverse or slanting cut, as described before. One cuts the scions, for which purpose he needs a
very sharp, thin-bladed knife ; a good budding knife is about
the best that can he had. A fourth inserts the scions, which
should be kept *moist*, not *wet*, by wrapping in a piece of moist
cloth ; the fifth ties, with raphia, basswood bark, or rice
straw, taking care to wrap the whole cut smoothly and evenly;
and the sixth covers up, first pressing a little moist earth on
the cut, and then filling up around the scion to the upper
bud, making a small mound around it. This division of
labor has the advantage that each hand has to perform only
one single and very simple operation, which he can soon
learn to do to perfection, and that even very common hands
can be used for the first and last operations. Nos. 2, 3, 4
and 5 ought to be the most careful hands, as is obvious, but
any good, handy man can soon learn these to perfection.

It will be easily seen how this can be varied by grafting
above or just at the surface. This will save the labor of digging around the vine, but increase the "mounding up"
around it. In this case I think that bandages and grafting
wax should be used, which one could apply with a brush after
the scion is inserted, over the surface of the stock, and over
the side on which the scion is inserted, covering the whole
place of union. If the one who ties follows closely, the
bandage will hold more firmly on the grafting wax. A very
good wax for the purpose is composed as follows : Two-
thirds rosin, one-sixth beeswax, one-sixth tallow, melted together and applied *warm*, not *hot*. It must not be hot
enough to burn the finger. A very convenient preparation
is made of shellac, dissolved in alcohol, say 1 lb. shellac in
a quart of alcohol. This can be applied cold, and is always
ready for use.

3. Grafting on non-resistant vines with the best of resistants, to establish the graft on its own roots.

To do this, we ought to graft as far below the surface as we can without injury to the vine, so as to secure its whole strength to push the graft, until established on its own roots. The only object I see in it here, is to multiply such varieties rapidly as are valuable, either for their fruit or for their wood for propagating. Of the first, there are but two which I should consider of sufficient value for wine making : the Herbemont and Rulander, also perhaps Louisiana, which has not been so fully tried, or a few of the Labruscas for table use, referred to before. For the latter purpose, the Cunningham and any other of the resistants, of which it may be desirable to grow a large amount of wood for propagation. Any strong and otherwise worthless Vinifera vines may be used as stocks, the ground dug away as deep as advisable, to just above the first whorl of strong roots, the vine sawed off, say 1½ inch above the node, and one, two, or even more scions inserted, according to the strength of the vine. The cut will heal over quicker when more scions are put on, and if some of them should be superfluous, they can be cut off afterwards. Cleft grafting will be most practical here, and the scions should be long enough that the upper bud is above the surface of the ground. No tying will be necessary. Then fill up with finely pulverized soil, pressing it firmly over and around the cut, to the top bud of the scions. If the scions have good, strong fruit buds, you can have the pleasure of tasting their fruit that summer. A Herbemont, grafted by me on a strong Isabella vine in spring of 1852, produced two canes of 26 feet, and one of 30 feet long, ripening nine perfect bunches of fruit besides, the same summer.

AFTER TREATMENT OF THE GRAFTS.

It will generally take some time before they begin to grow, as a firm union must take place first, before the scion will be in a condition to grow vigorously. Do not get discouraged if the majority of the scions do not show any progress for a month, for those which start so early generally wilt back and do not amount to much. I have had grafts to start in August, and make a very satisfactory growth. As long as the scion is fresh and green, it shows that it receives sustenance from the stock, and may start at any time. They should be examined from time to time, as the suckers from the old stocks may intefere with their growth, and just as soon as these appear they should be removed, by digging down to the place from where they start and taken off clean, for if any part of them remains they will throw up four or five in place of one. The suckering should be done twice, at least, taking care, however, not to disturb the grafts; once at about a month's time from grafting, and the other at about two months; the remainder, if any appear, may be pulled up easily at any time.

As soon as the young growth on the grafts appears, they should be staked, putting the stake on the side from which the prevailing wind blows. As the young shoots are very succulent and tender, they are very easily blown off by the wind, and should be tied as soon as six inches to a foot high. If the stocks are strong, they will grow with astonishing rapidity, and twelve to fifteen feet of growth is very common. I generally leave all their laterals and pinch off the leader when it has reached the top of the stake, which will make them grow more stocky. If on strong vines, they will generally be ready to bear nearly a full crop the following summer, so that there will be little, if any, loss of time. They should not be tied too close, so as to give them room to expand, only to offer the proper resistance to the wind, with some soft material, either with strips of the Phormium Tenax (New Zealand Flax) recommended and distributed by Prof. Hilgard, or Dracæna and Yucca leaves split into strips, of which I shall speak more at length, in "tying the vine."

In France, there is a good deal of grafting done in the shops, or in nursery. I have tried repeatedly to graft young vines in the shop in winter, as is done so much with fruit trees, but never had success enough to induce me to recommend it. It seems that they need a strong and vigorous flow of sap, to unite well, and that this is too tardy in the transplanted vine to make a good union. If cuttings or seedlings are planted in nursery, not too close, and grafted the following spring, this may do very well. They ought, in that case, remain in the nursery until next spring, or winter, and then transplanted at once to the vineyard. This would be valuable to filll vacancies especially. But on the whole I much prefer grafting in vineyard. The cultivation of the grafts will be similar to that of other vines, and therefore need not be discussed here.

HERBACEOUS OR GREEN GRAFTING.

This is a good deal practiced in Hungary, and is strongly recommended by that eminent practical grape grower, and writer on viticultural matters, Herman Goethe, Director of the œnological institute at Marburg, to whom we owe some of the best books on Viticulture we possess in the German language. It is rather a supplement to spring grafting, and would be of great value in many respects, if it can be made successful, also for grafting in nursery during the summer. It would be easy in this case, to graft the suckers of such vines as may have failed to grow, and thus make the stand complete the first year. But after a trial of two seasons, I fear it will not prove successful here, as very few of the grafts united, but nearly all wilted off. I think that our summers are too dry at the period at which it ought to be performed, in June, to make it practicable. This may, however, be overcome in a certain measure by grafting close to the surface, filling around the junction with earth, to the upper bud of the scion, and then watering it, so as to keep it moist. For the benefit of those who wish to try it, I will describe and illustrate it here, quoting from Prof. Goethe's valuable Pamphlet "die Reben-verdelung" (Grafting of the Grape).

"For over fifty years a method of grafting has been practiced near Funfirchen, in Hungary, and also in other districts, which has been called *green* or herbaceous grafting, because the soft green shoots of the vine are used for the operation. It was practiced mostly to change single vines in vineyards that were of an indifferent variety, into such desirable varieties as composed the greater part, or to change whole vineyards into better varieties. But it may now become of still greater importance, if we apply it to protect our European varieties, by grafting on resistant American stocks, against the attacks of the phylloxera, and it has already been successfully used for that purpose.

"Experience in grafting all kinds of plants shows us that the operation is followed with a greater degree of success, when the parts to be united are yet young and succulent, than when they have already become woody and hardened; always provided that they have the necessary firmness and development to be capable of germination. This is also applicable to the vine, and if we have to record a good many failures, the causes of them were mostly either due to improper time or faulty performance of the operation itself.

"Although the operation was fully described in the work of Franz Schams, perhaps the best informed writer on the culture of the grape in Hungary, which appeared in 1832, and although it has also been practiced in other districts of Hungary, for instance in the vineyards around Ofen, where I saw its application, yet it has not, to the best of my knowledge, been introduced into other provinces. This may be owing to the fact that most of the experiments made with it did not give favorable results, and therefore I applied myself to find the cause of these failures, and to learn the proper method to avoid them.   I think that all are easily explained, and the success of the operation depends mostly on a few very simple points, which, however, are of the greatest importance.   If performed correctly and at the proper time, nearly all the green grafts will grow, in Hungary as well as in Marburg, and even the grafts made by our students have taken, with only a loss of from two to three per cent.

" As I have remarked before, the *proper time* is very material.   In districts which are subject to late frosts, the work cannot be performed until all danger from them is past.   Therefore we will presume that for most districts the proper time would be from end of May to beginning of July.   But the best guide in this respect is the condition of the young shoots.   The stocks as well as the scions must be *elastic* and *pliable*, not too woody, nor yet too soft.   The pith of the young wood must be yet green, not show the white tinge which it assumes later.   Of course, the buds must be sufficiently developed to grow.   This is generally the case at the last fully developed leaf.   This will explain why all shoots on a vine or all plants in a nursery are not in a proper condition at *the same time*, and that only those should be selected which are sufficiently developed and growing rapidly.   Laterals with points that have already completed their growth are unfit for use.   It is of the greatest importance that the stock, where it is to be grafted, is still growing and juicy, so that we may expect a vigorous growth.

" It has also been observed that sunny and warm weather, when the vine grows rapidly, is more suited than cloudy and cool days, when the vine is not so thrifty.   The operation should not be performed in the hottest hours of the day, therefore it will not succeed well during continued dry weather.

"The operation itself is very simple, as it is very similar to common cleft grafting. The scion is cut to a short wedge at a node, so that the point of the wedge is yet at the firmer part of the node, and the upper part cut off about half an inch above the bud which is expected to grow. The scions thus prepared are either used at once, which is perhaps best, or kept in water with their lower end, but the water should be shaken off before insertion. The leaf opposite the bud is cut back to the stem.

"The stock should be in about the same condition as the scion, but may be somewhat larger than it. Cut it close below an eye or bud, take away the young laterals below, and split it exactly in the centre down to the next node or bud, so that the split goes down into the firmer structure of the node.

"The scion is now inserted so that the outer green bark is even with that of the stock, and the wedge fits closely into the end of the split. The cuts must not be bruised in insertion, which is easily prevented by holding the split slightly apart. As soon as the scion is properly inserted, the split is firmly tied with cotton twine, especially when the lower end of the scion is joined to the notch of the stock. The best plan is to commence tying from above, so as to prevent the scion from slipping back. See Cut, Fig. 10.

"After six to eight days, the scion will show very plainly whether a junction has been made, as in case of failure it is dried or withered. If the scion has taken, the leaf stem opposite the bud has dropped off. As soon as the scion begins to grow, and the thread cuts into the bark, the tie should be loosened.

"When herbaceous grafting is applied to the vines, it will be advisable to have but one shoot on each arm, and take off all the others; if the vine has several arms, one ought to be left on each of them. Of course it is important to give

the grafts support, if
they have taken, by
tying to stakes, and
they can be bent
down later."

(I hardly think that
this grafting above
ground will succeed
in our climate, unless
the canes or shoots
are bent into trench-
es and covered with
earth.—(G. H.)

"This process
seems of more im-
portance to me, how-
ever, when applied
to young vines in
nursery, as has been
done at Marburg in
1878.   In this case,
we take young Am-
erican vines as stocks
and graft with Eu-
ropean varieties.
These can be remov-
ed to the vineyard the
following season, and
we would thus have
the most suitable
vines for planting
on resistant roots.

" But whether ap-
plied to old vines or
in the nursery, it is

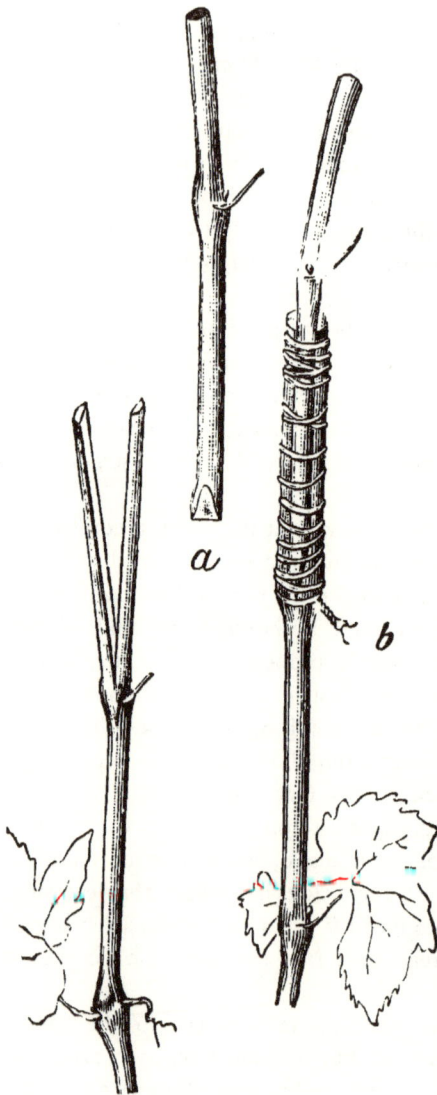

a   *b*

FIG. 10.

very important to take off all shoots which may come from the stock below as soon as they appear, and we have evidence that the graft has taken; so that the entire strength of the vine is imparted to the scion."

This comprises about the most important information from Professor Goethe, in relation to green or herbaceous grafting. As mentioned before, my experience with it so far has been discouraging, and I fear that our summer weather is altogether too dry. But its advantages would be so manifold, that it well deserves a thorough trial.

Mr. Clarence J. Wetmore has tried another method of grafting in August, with which he claims to have had uniform success. We know that the vines make a second growth in August, which may be sufficient to effect a junction between stock and scion. The vine is grafted in the usual manner, below ground, the scion taken from the young but well ripened wood of this year, and Mr. Wetmore claims that he loses hardly any, although the scions seldom grow much the same season. They merely make a junction with the stock, but grow readily and vigorously the next season. This method is also worth trying, where the scions may have failed in spring.

These comprise about all the methods of grafting which are practicable here, and we can close this very important subject with them. It would only serve to confuse the reader to name and describe the innumerable varieties of these and others, and be of no practical benefit.

Budding has often been tried, but as far as I know with little or no success; therefore I omit it altogether.

In this, preeminently the "Golden" State it is said that we have more of first class grape lands than all of Europe put together. This may be true, but it is also true that but few of the best lands, those that will make a name and fame for our industry, have so far been planted in grapes. Those lands that could be planted to vines with the least labor, were the most easy of cultivation, and produced the largest crops; were generally chosen and planted. While I do not undervalue easy cultivation, I also believe that our rich, level bottom lands will never give us the wines that will rank with the finest brands of Europe, or even surpass them. All we can expect of them is a good, sound wine for the million, the every day drink of our people, and while I admit that this is a great desideratum, yet the small grower especially should aim at higher quality, which will make his wines and grapes sought for at high prices.

Then again, our lands are very variable, they are "spotty" as we familiarly call it, to a very high degree. It is seldom that a piece of one hundred acres, or even ten to twenty, can be found which is alike, or only nearly so. The soil is apt to vary from deep, naturally drained land to hard-pan alkali, from stony and pebbly soil to stiff adobe or clay, and again to shallow bedrock, where there is not depth of soil to let the vine root fairly, and develop fully. It behooves every one then to be careful in the soil he selects, and to look over it well before he plants it to vines.

The location is also a very important question. It should be easy of access, should have fair transportation facilities by railroad, or, at least, a promise of them in the near future, for this is a question which will influence the value of the product very materially. Then the aspect of the place itself, its features, whether they allow easy communication and centralization, should be well considered. The vineyard should in all its parts be easily accessible to either the wine cellar or the packing shed. If these can be so located that all the grapes can be brought down to them, or at least on a level, it will make all the operations a great deal easier. And for the wine cellar especially water facilities are a great consideration. If living water from streams and springs is available, that can be led into every story of the cellar and into every compartment, it is an advantage which can hardly be overvalued. It is not alone handy for cleansing casks and vats, in short, the whole working apparatus, but it makes the work itself so much more convenient and so much less costly. Besides, it adds greatly to the coolness and cleanliness of the place. If good material for building is at hand it is also a great advantage. A hillside for the cellar, into which it can be excavated so that every story of the building can be approached by wagon, is a great advantage, and if good building stone is at hand close to it, or at the cellar, it is worth a great deal also. All these points should be carefully considered before the land is selected. They cut a very large figure in the expense account. It is fortunate, indeed, that our mountains are blessed with so many never failing springs of the purest water, which are available for the purpose; and that good building stones are also by no means scarce. They are one of the most attractive features of our noble State.

When there is no spring or stream available, there should at least be a good well, from which the necessary supply can be pumped by machinery. A wine cellar without a full supply of water is a very poor institution indeed.

There is another and serious objection, at least in the more

northern part of the State, their greater liability to late spring
and early fall frosts. It is rather unpleasant, as well as laborious, costly, and fatiguing to watch the thermometer every
cool night, to see if it does not approach the fatal 32°, and
although I hope to show in a subsequent chapter that frosts
need not and ought not to be as destructive as is now supposed, yet it is a very discouraging thing so see a vineyard,
beautiful in promise but the day before, blackened and wilting
before that invidious enemy in the morning. It is well known
that an elevation of a few feet is often sufficient to save the crop;
therefore gently sloping hillsides should always be preferred
to valley lands.

As to the particular aspect, this is not so important in this
State, where all grapes ripen well enough, and are rich enough
in sugar. The northern and northwestern slopes, however,
when not too steep, generally have the deepest and richest
soil, though there are exceptions even to this rule. The timber growing on the land is generally a good indication of its
adaptation to vines. Where there are large and heavy oaks,
manzanita and madrona, where the poison oak (rhus toxico-
dendron) grows luxuriantly, interspersed by the hillside fern,
and also in the red soil of the redwood region, where this noble tree, the fir and Douglass spruce flourish, the soil is well
adapted to the vine, though in the latter region also difficult
to clear, an item which should also be taken into consideration.

Chemical analysis of the soil, of course, will show us
whether that particular soil is desirable. But the difficulty
has been mentioned before under which we labor here. Our
soil is so diversified that a chemical analysis, unless made of
every acre or two, will not give us anything positive to stand
upon for a large tract, however sure it may be for the partic-
lar spot from which the soil was taken.

I prefer the soil to be light and friable, although I have

seen fine vineyards and excellent wines from tenacious clay or adobe soils. Still such soils are very difficult to work, as they bake and roll when wet, and get very hard when dry. If the soil is free from stones, so much the better, for stones make it disagreeable to work, although stony soil, if otherwise rich, will make fine wines. But avoid hardpan and alkaline soils, as they are not suited to the vines, will produce but little, and that little of inferior quality. It should not be too steep, as that makes cultivation difficult and costly.

With these general hints, I will shortly sum up the requirements of as nearly perfect a vineyard as I can imagine.

1. Easy access to market by railroad or water, and from vineyard to cellar.

2. Gently sloping lands, not too steep nor too flat.

3. Abundance of water.

4. Good location for cellar and packing sheds, and, if possible, good building stone.

5. Freedom from frosts.

6. Deep and friable soil, which, for red wines especially, should be rich in iron and other minerals, to give color and tannin.

CHAPTER X.

PREPARING THE SOIL.

This is very important, especially in tenacious soils, clay and adobe, which should be deeply stirred, to enable the roots to penetrate it. Where the soil is naturally loose, it is not so imperative, although it is always well to have it thorooghly mixed and pulverized.

Of course, I take for granted that the soil has been cleared of all roots, stumps, stones, etc., before plowing. When the soil has been thoroughly moistened by rain, so as to work satisfactorily without being wet, put in a good team with a strong turning plow, which can make a furrow of from 10 to 12 inches deep. If two horses cannot do it, put in four, and follow in the same furrow with a subsoil stirring plow, that will only loosen the soil, not turn it. This ought to stir it from 4 to 6 inches deeper, so that the whole depth is from 14 to 16 inches at least. Of course it depends on the nature of the soil how many animals ought to be used, as it will be much harder of draft in tenacious, clayey soil, than in loose and friable earth. Where stones, roots, etc., are turned up, they ought to be piled up and taken out; and it will be well to employ an extra man for that purpose, who can follow the plows, and remove any obstacle they meet with.

After thus thoroughly loosening the soil, it should be harrowed crosswise, and then gone over with a clod-crusher or drag, to leave an entirely smooth and even surface. Remember that you are laying the foundation for work which is expected to last your lifetime and longer, and rather spend five or ten dollars worth more labor to do it well, than to do it poorly.

Wet spots should be drained by gutters, either of tiles or stones. Make a ditch 3½ feet deep by at least a foot wide, setting two stones on edge, then laying a flat one on top; then throw a layer of straw over these and some loose stones on top, filling up with soil. These I have found to carry off the water better than drain tiles, and where stones are convenient, they are much cheaper, and help to clear the land, as it will take a good many stones, which will be buried under ground, instead of being obstructions in cultivating.

CHAPTER XI.

HOW TO LAY OUT AND PLANT A VINEYARD.

Every vineyard ought to have a main road or avenue, into which all others lead, and which should therefore, if the location will permit, be as near central as possible. If the ground is rolling, and cut up by ravines, the greatest convenience in hauling and carrying will have to be consulted. In ground nearly level, or gentle slopes, the most convenient and economical plan will be to lay it off in squares of about two acres each, but making the squares double as long as they are broad, to facilitate the carrying of the grapes, stakes, fertilizers, if necessary, etc. The best distance between the rows I consider 8 feet, though many plant 7 feet, and even closer, while others contend that they should be still further apart. Eight feet will give room for convenient working, and also give sufficient space to the roots. It will take blocks of 25 rows broad and 50 rows long to make about two acres of vines, but if we want to include the avenues in the measurement, the block should be 24 by 48. The first will give 1246, leaving off the four vines at the corners, for greater convenience in turning, the latter 1148 vines to each block.

The next thing we want is a lot of markers; that is, short, thin stakes, split of redwood, say 15 to 20 inches long, and about half an inch in diameter. They are tied in bundles of 100 to 200, and to make them more apparent, the tops can be dipped in a tub of whitewash. This, with two long lines as long as a block at least, or 400 feet, and a short one to reach across the block, or 200 feet, two measuring poles

8 feet long, and four men, complete our equipment for the simplest and most expedient plan for laying off I have yet tried, and which will always give straight lines, if the avenues are correctly marked.

We commence by establishing the main avenue or road, first running a straight line through the center, if our piece of ground admits it, in dimension and nature of soil. This we do by setting a few long stakes or guides, then drawing our line from one to the other, one of our men having hold at each end, and he takes a measuring pole eight feet, or any other desired distance, the fourth carries a lot of markers, putting one down every time the distance is measured, and counting. When twenty-five are counted, the breadth of a block, omit one, measuring sixteen feet to the next, so on to the end of the field. We now run another line, parallel with this, but sixteen feet from it, in the same way, and this establishes our main road or avenue.

We now measure in the same way around each block, taking care to have them at right angles. If the main avenue runs north and south, we run our next line below east and west, or the reverse, and count fifty ; from there to north and south again, counting twenty-five, and back to the main avenue and the 25th marker. It is best to establish all the outside lines of the blocks first. When we get this done, the rest will be comparatively easy. We now draw a long line from marker 2 longitudinally, over the block to marker 2 on the other end, also the other long line from marker 24 to 24 at the other end. Two of our men now take the short line on each side of the block, holding them to the next marker east, while the other two take each a bundle of markers, and put them down in the angle of the crossing lines. If the men move lively and precisely, and take care to stretch their line well, it can be done very fast and accurately. As soon as the marker is down, they both move the line to the next, and

so on until to the end of the block. The long lines are then
moved to Nos. 3 and 23, and the same repeated until the
block is finished. The intelligent reader will easily see how
this plan can be varied, according to circumstances, distance
to be planted, location, etc. If we have rooted vines, we
make holes next, to receive them. This is best done with a
spade, putting the spade down just above the marker, as nearly perpendicular as can be, taking out its full depth, say
twelve inches, and putting the ground taken out on the upper
side of the hole. The planter follows with the vines in a pail
partly filled with water, to keep the roots fresh. The roots
should first be shortened into a uniform length with knife or
shears, and if resistant plants are used, which are to be grafted
afterwards, the lower buds along the stem should first be
cut out, as mentioned before, to prevent suckering.   In
planting, spread the roots evenly at the bottom of the hole,
giving them a downward position, (Fig. 11), then fill up with
well pulverized, moist soil, which may be pressed down, but
not roughly tramped, with the foot. The top or head of the
young vine should be even, or a little above, the surface of
the ground, and come out close to the marker.

Any time during the winter is a good season for planting,
but the ground should work well, be neither too wet nor too
dry ; and if planted early and heavy rains follow, the ground
around the young vine should be stirred and made mellow in
early spring. I do not think there is much gained by very
early planting, as the ground is too cold then, to give the
young vine much of a start. But we are often crowded with
other work in early spring, and the vinegrower, to be successful, should always rather be ahead than behind with his work,
and should take advantage of every spell of fair weather that
comes to his aid.

If cuttings are planted, it may be done in the same way,
only I would advise taking two for each hole, to avoid

vacancies.   They can be placed about six inches apart with
the lower end, while the tops come together at the marker, so
that one can be removed, should both grow.   I do not think
it advisable, however, to plant cuttings *early* in the season.   I

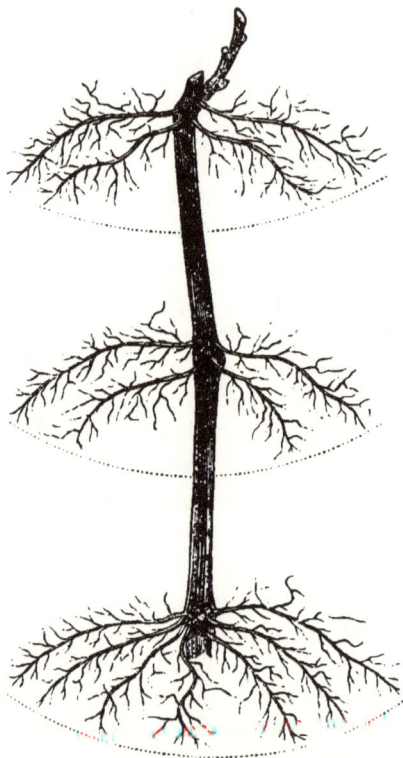

FIG. 11.—YOUNG VINE READY TO PLANT,

would rather keep them heeled in, reversed, as is described in
the chapter on propagation, and plant when the soil is somewhat warm already, not before March or April.   They will
callus and root quicker than if they are put into the cold
ground in winter.

CHAPTER XII.

WHAT TO PLANT. CHOICE OF VARIETIES.

I need not repeat here, what has already been said about resistant roots in a former chapter. I would not plant any other, even in locations not yet infested with phylloxera. Of them, the Herbemont, Rulander, Louisiana, and perhaps Lenoir may be used for direct production. All the others should be grafted with the best foreign or Vinifera varieties ; but although that will not be done until a year later, we may as well consider the question here. Which *are* the best ?

This is a knotty problem to solve, in a State where there are collections of from three to four hundred varieties, which pass under one name in one locality, and in the next vineyard or valley perhaps, are grown under another. There is an almost endless confusion in this respect, and it needs great caution to get any variety pure and true to name. Yet there is nothing more vexing than such blunders, and I would rather have the money stolen out of my pocket by a thief, than to be thus swindled, whether intentionally or not. Therefore, be sure of what you obtain, and get it only of reliable men. One of these is Mr. H. W. Crabb, of Oakville, Napa Co. He has a collection of nearly four hundred varieties, and spares no pains to have them correct. Better pay treble the price, it amounts to but very little anyway, and get them true to name. The next consideration is, "what do we want them for, for *wine*, table or market fruit, or for raisins ?

If for *wine*, that again depends upon locality and soil. It is of no use to try and grow grapes for red wine on soil that will not give us color, astringency or tannin, and fine bouquet.

To do this, it should be rich in minerals, in iron, especially ; and as enough of red wine grapes are planted already on soil not adapted to them, do not let us add still more to them. Nor let us plant any, red or white wine varieties, in soils and locations where they are inclined to turn into sherry and port. To produce fine light or dry wines, I think we will have to confine ourselves to Northern California, or to those elevated regions in the Southern counties, where grapes will not ripen before September, and we can take the fall months, September, October and even November, to give them such gradual and moderate fermentation, as will enable them to attain that fine bouquet which alone will make them of permanent value in the markets of the world.   Let each grower confine himself to his proper sphere, taking advantage of the indications which his surroundings give him, and make such a product as he can make in the greatest perfection, be this light wine, or the heavy ports, sherries, and sweet wines, or brandies.

Light dry wines being used in the largest quantities, we will consider them first.   To make them, we want grapes that will give.

   1.   Fine *quality*.

   2.   Sufficient *quantity* to pay well.

   3.   Varieties which are easy in cultivation and training, or in other words, which will give the *best* returns for the *least* labor.

   4.   Varieties *easy to handle and ferment in the wine cellar*, and which are most in demand, commanding a *ready sale*.

With these points in view, we will now consider the varieties best suited to " fill the bill."

FOR LIGHT, DRY, WHITE WINE.

Pedro Ximenes.   Synonyms, Sauvignon Vert, White Green Riesling, Columbar erroneously.   This is not a Riesling, but one of the Sauterne type of grapes.   It is a very strong, vigorous grower, a good bearer with moderately long pruning, very healthy, also suffering less from frost and coulure, and will become, considering all this, one of the leading varieties. Wood brownish gray, dotted with darker spots, rather long jointed; buds whitish, prominent. Leaf dark green above, somewhat rough, light green below, covered with gray hairs, stem of leaf brownish, points of young shoots gray and hairy, tendrils strong, generally divided into three at the end. Bunch long, rather loose, shouldered, with long medium sized stems; berry medium, slightly oval, greenish yellow, translucent; thin skin, very juicy, sweet and sprightly. This grape makes a very sprightly, high flavored and smooth wine, which can bear diluting one-fourth in drinking, and still retain its full character. It has a fine persistent foliage, and will not suffer from sunscald. Ripens here in Northern California about the end of September.

Marsanne. Synonym, Avilloran. This is another of the Sauterne varieties, but of a stronger flavor than the former, therefore it should be blended with lighter varieties, such as Burger, for instance. Vine a very vigorous grower and immense bearer, can be grown on four foot stakes with short pruning. Wood dark gray with brown spots, growing nearly straight, rather long jointed. Leaf dark green, rough on upper side, grayish green below, leafstem long and stout, green, young points of shoots, gray and wooly, tendrils long, forking into 3 to 5 points. Bunch large and heavy, shouldered, moderately compact; stem thick and long; berry rather small, round, yellow, covered with white bloom, and when fully ripe has a brown tinge in the sun; moderately juicy, rather thick skin, sweet and high flavored. A very healthy vine, but so productive that it is apt to overbear. Foliage fine and persistent, never suffers from sunscald, ripens rather late, middle of October here.

Green Hungarian. Synonyms, Verte longue, Long Green.

The origin of this grape is somewhat doubtful, but not its great value as a wine grape. I received it from Mr. Groezinger, under the name of Green Hungarian, have seen it bear for four consecutive seasons, and think it a model vine in every respect. Immensely productive, a short and stocky, but vigorous grower, splendid foliage, and easy to handle, it " fills the bill " more completely than any other grape I know. Its wine comes nearer to the Riesling type than the Sauterne, is sprightly, high flavored, greenish yellow, and with the pleasant piquant acidity of the Riesling, while it will bear three times as much.

Wood grayish brown, short jointed, vigorous. Young growth stocky, green with brownish veins, furrowed. Leaf heart shaped, but slightly lobed, sharply but irregularly serrated, full as broad as long, light green above, pale green below, covered with fine hairs, young points gray and tomentose or wooly, tendrils rather thin, with only one fork. Bunch long and heavy, sometimes weighing three pounds, shouldered, compact, with a stout but rather short stem ; the shoulder often nearly as heavy and long as the main bunch, which last often has a double point. Berry small, round, but often pressed out of shape, as they are so close on the bunch, greenish yellow, covered with white bloom, moderately juicy, very spicy and agreeable. Ripens here about last of September.

I have been thus minute in describing it, as it seems to be very little known, certainly not half as much as it ought to be. I think it will produce some of our finest wines, and is fruitful from every bud. Can be grown on short stakes, and pruned to spurs, and has produced for me forty pounds to the vine, on Riparia roots, the second season after grafting.

Chauché Gris. Synonyms, Gray D'Jshia, Greg Riesling. This is not a Riesling, but only called so erroneously in Napa Valley, where formerly any grape with small compact bunches and small berries was classed with the Rieslings. It belongs to the Burgundy or Klaevner type, and is very similar to, but not identical with, the German Rulander or Grey Klaevner, It is a stronger grower and more productive, with larger bunches and berries, though not of as high quality. However, it is a truly fine grape, an abundant bearer with long pruning, though sometimes subject to coulure, and makes a very fine wine, if not allowed to get over ripe, or fermented with some grape of lighter quality and a more sprightly acid.

Vine a strong upright grower, with straight brown wood. Leaf dark green above, lighter green, somewhat downy below, medium size, deeply lobed, young shoots green, points of shoots grey, tinged and edged with carmine, tendrils slender, three pronged. Bunch compact, mostly shouldered, short, strong stem; berry small to medium, pale red or grayish, covered with gray bloom, slightly oblong; skin rather thick, moderately juicy, high flavor. It ripens suddenly about the end of September, and should be taken when not higher than 25^{C} Balling, when it will make a much more sprightly wine than when over ripe. In the latter case it is sluggish in fermentation and apt to give trouble, unless fermented with some lighter variety.

Semillion. Synonym Colombar. This is one of the celebrated French varieties, a combination of it, the Sauvignon blanc and Muscadelle de Bordelais make the famous wine of Chateau Yquem. It seems to do well here, with long pruning, and certainly makes fine wine by itself. Wood grayish, strong and straight, leaf medium size, downy, lobed, wavey and irregular. Bunch medium, seldom shouldered, but broad, moderately compact ; berry medium, slightly oblong, greenish yellow, very thin skin, transparent, very juicy and sprightly, high flavor. Young shoots green, with grayish points. Ripens here the first week in October. and makes a very delicate, high flavored, and sprightly wine, of greenish yellow color.

Tramenir. Synonym, rother Klaevner. This is not a very productive variety, although a fair bearer, but it brings pretty sure crops, and the wine is of such superior quality, which it will impart to other lighter wines, that it ought to be in every vineyard, if only a few acres. Vine a moderate grower, wood short jointed, thin, grayish brown, changing to ashy gray, hairy, buds whitish gray. Leaves small, round, thin, often broader than long, dark green above, light green below, hairy and downy, stem of leaf reddish, points of young shoots grayish white, with very small leaves. Bunch small, compact, sometimes shouldered; stem short and brown; berry small, oblong, or oval, pale red with gray bloom, skin thick, moderately juicy and very sweet and spicy, ripens about end of September here. Like the Chauche Gris, to which it is closely related, but superior in quality, it ought not to hang until over ripe. The noblest wine I have seen in this State yet was made from this grape.

Sultana. This is perhaps not suited to all localities, and should not be planted when there is danger from late spring frosts, as the vine starts early. Yet it makes, in northern California, a very fine wine, and as it is a very abundant bearer, with long pruning, and the berries are seedless, it is a very profitable grape, as it can also be used for choice raisins. Vine is a strong grower, brown, long jointed wood. Leaf, thin, bright green above, lighter green below, smooth and shining, deeply lobed and sharply serrated; young wood dark green, points greyish brown, tendrils at every joint, thin and slender. Bunch very large and loose, shouldered; berry small, round, golden yellow, covered with light bloom, sweet and juicy, firm and crackling, without seeds. Ripens here end of September, and makes a very delicate wine of straw color, great body, and acquiring with age a natural sherry flavor.

Riesling. True Riesling, Johannisberg Riesling. This and the following are hardly productive enough to be classed with foregoing varieties, but we can hardly omit them, as their superior qualities for wine make them indispensable in every vineyard. The Riesling has given to the Rhenish wines their high reputation for delicacy, sprightliness, and the high bouquet which has made them known and famed throughout the civilized world. Vine moderately vigorous, wood straight, light brown, speckled with white and darker spots, short jointed. Leaf rather small, round, thick and rough, deeply lobed, grayish green above, light green below, with yellow spots here and there; leaf stem thick, reddish, with rough warts, points of shoots and small leaves yellowish green, wooly, with faint reddish tinge. Bunch small, compact, short thick stem. Berry small, round, light yellow, with black dots, transparent ; when fully ripe tinged with brown ; skin thick, juice sweet, very aromatic and high flavored, ripens first of October.

This vine needs long stakes or trellis, and long pruning to canes, and is then apt to lose a large part of its crop by coulure, or imperfect setting. It is therefore not a very profitable grape, but like the Traminer, deserves a place in every vineyard to make a superior product.

Franken Riesling. Sylvaner, Oesterreicher. This is a somewhat better bearer than the foregoing, though it is also subject to coulure, and a difficult vine to handle, on account of its strong and bushy growth. Yet it makes a very fine, smooth, and agreeable wine, of fine bouquet, though not equal in that respect to the true Riessling.

Vine vigorous, close jointed, and bushy; wood, light brown, with darker spots; buds, small, brown; leaf medium, round, slightly lobed, thin; bright green and shining, light green and smooth below, with yellow spots in fall; stem of leaf, short, thick, with reddish tinge, points of shoots bright green.

10

Bunch small to medium, very compact, sometimes shouldered; stem very short and thick, which makes it difficult to pick. Berry round, though often pressed flat by being so close on the bunch, yellowish green, with a small dot, medium in size, covered with thin white bloom, skin thick, juice very sweet and spicy; ripens end of September. It also needs long stakes and long pruning to canes to bring good crops.

Muscadelle de Bordelais. Synonyms, Musquette, Raisinote, Cadillac. This grape promises to be of great value here, on account of its peculiar, spicy flavor, which is used to give the fine bouquet to the celebrated Chateau Yquem wine. It is a very strong, stocky, robust growing vine, and seems to be productive. Wood, brown, short jointed and stocky; leaf dark bright green above, shining, paler green below, nearly round, slightly lobed; points of young shoots and leaves brownish gray, tendrils simply forked. Bunch small to medium, shouldered, compact, short, thick stem; berry slightly oval, light yellow. thin skin, very juicy and sweet, with a very delicate, spicy and aromatic flavor. It is here hardly long enough to be fully tried, but certainly deserves a place in every vineyard where quality is an object.

Clarette Blanche. Synonyms, Granolata, Blanquette de Limoux. This is also one of the recent introductions, but so far has proven itself a very strong grower, abundant bearer, and making a wine of superior quality. I would not yet advise its general culture, but recommend it for trial to those who have long seasons to ripen, and strong soil. Vine a strong grower, wood brownish, short jointed. Leaf large, rough, dark green above, grayish white below, woolly, points of young shoots whitish gray, very downy, tendrils small, forked. Bunch medium, broadly shouldered, moderately compact, stem thick and short. Berry oblong or oval, medium, greenish white, very juicy, skin thin. The samples I have seen of the wine, were very sprightly and delicate.

It ripens late, about the middle of October in Napa, and would therefore not be suited to northern localities, while well adapted further south.

These are all high quality grapes, most of them very abundant bearers and will certainly afford choice enough in white wines for any vineyard. There are of couse many others, which are very promising, but most of them have not been sufficiently tried to recommend them. I will add to these a few varieties of not as high quality, but very productive, and which may be advantageous to blend with some of the foregoing.

Chasselas Fontainebleau. Synonyms, Golden Chasselas, Sweetwater, Gutedel. This grape has been successful nearly everywhere, is a good and regular bearer, and makes a very fair wine, although not of very high character. It is one of those vines which will not disappoint the planter, and which can be much improved by blending with higher grade varieties such as Marsanne, Green Hungarian and Traminer. Vine a strong but slender grower, which will bear well with short or long pruning, low or high stakes. Wood brown, long jointed, slender; leaf thin, deeply lobed, bright green, ends of shoots and young leaves brownish green. Bunch medium, compact, shouldered; berry medium, round, yellow, transparent, of a peculiar crackling firmness, juicy, sweet, but without any very high character. Ripens early, about the middle of September here, and is also one of the earliest market grapes.

Chasselas Violet. Synonym, Kœnigs Gutedel, Violet Chasselas. This is a grape of higher character than the preceding, but must be pressed soon after crushing, or the color of the wine will be too dark, turning a reddish yellow tinge. Also a very reliable bearer. Vine a strong, long jointed grower, wood dark brown, with a violet tinge; leaf thin, dark brownish green, deeply lobed, young growth brownish.

Bunch long, shouldered, compact; berry round, pale violet
red, and has the peculiarity of acquiring a violet tinge when
only half grown, sweet and juicy. Makes a very agreeable
wine, and ripens at same time with the foregoing.

Victoria Chasselas. Queen Victoria. This is in my opinion the most valuable of the Chasselas family, as it really
makes a fine wine, is easy of culture, and a great bearer.
Vine vigorous, very short jointed and brittle, wood grayish
yellow, thick and strong. Leaf bright green, deeply lobed
and shining, young shoots with numerous laterals. Bunch
very large and heavy, often weighing five pounds, shouldered, very compact, stem brown, very thick. Berry
medium, round, pale lilac purple, with lilac bloom, juicy,
vinous, and refreshing. This is a fine grape, easy of culture, as it will bear well with short pruning, easily picked,
and deserves more attention than it has received so far.

White Elben. Synonym Elbling, Kleinberger, Kleinbeeriger.
This variety is cultivated considerably in Sonoma Valley,
where it bears fairly well, often very abundantly and is
prized for the lively wine it makes, which though not of high
character, has a very agreeable lively acid and pleasant bouquet. 1 would also think it a fine wine for blending with Traminer, Chauche Gris or Marsanne, as it will serve to relieve
the abundance of their flavor, and lack of tartaric acid. Vine
a strong grower, wood brown, with black spots. Leaf large,
rough, heart shaped, seldom lobed, dark green above, light
green and woolly below, leaf stem short and thick, hairy, reddish, young points of branches reddish green. Bunch large,
shouldered; berry rather large, round, but the vine has the
peculiarity, unless the blooming season is very favorable, to
set imperfectly, and thus a great many small berries are scattered among the large ones, which are very sweet and fine,
but only one fourth common size. This has given the grape
the name " Kleinbeeriger," by which it is known in many

parts of Germany. It requires high stakes and long pruning, ripe about last week in September.

The Blaue Elben, introduced and cultivated largely by Mr. L. J. Rose of San Gabriel, and of which he has made some very fine wine, is similar to its white sister except in color. But although the grape is black, its juice and skins contain very little color and tannin, and it is used for white wine. There is but little doubt that these lighter varieties, the White and Blue Elben, Burger, Folle Blanche, etc., may be better adapted for wine at the south than they are here, and prove *the* varieties for them to plant, if their aim is to make dry, light white wines.

Burger. Synonyms, Putzscheer, Large White Tokay. This grape has become so universally known under the name of Burger in this State, that it will be impossible to change it, although Burger is only a local synonym for the White Elben in Germany. It is here, in Northern California, regarded as only a quantity grape, for it is certainly incredibly productive, but in the valley lands it yields an entirely neutral wine, thin, acid, and without character. On rich hillsides its product is somewhat better, and in good seasons it makes a very fair wine. In the South, however, it improves and may be valuable for light wine; certainly Mr. Rose has made creditable wine of it at San Gabriel. However, its foliage is not very good, and its fruit therefore liable to sunscald. I do not wish to recommend it for extensive planting, but where it is already planted, it may be used to good advantage as a blend with very heavy, rich wines. One fourth Burger, added to these, often gives them sprightliness, and just the agreeable acidity they may need. Besides, it always ferments well, and is of value as an addition to such varieties in the fermenting vat, as are sluggish in their fermentation.

Vine a strong grower, with dark brown wood, speckled with black. Leaf nearly round, but lobed, light green above

with white down beneath, stem of leaf reddish, young points
of shoots white and wooly.　Bunch very large, loose, shouldered; berry round, of somewhat unequal size, whitish green,
covered with white bloom, skin thin, very juicy but watery
and thin.　It need hardly be mentioned here, that this is not
the grape which makes the celebrated Tokayes wines.　These
are made from the Furmint or Yellow Mosler, an entirely different grape.　The Burger ripens late, about middle of
October.

Folle Blanche.　La Folle, Enrageat.　This is called "the
crazy" on account of its heavy bearing in France.　It seems
to be somewhat like Burger in that respect, and of a similar
character, and may play a similar role as a desirable addition
to very heavy musts, and for fine brandies, for which purpose
it is used in France to a great extent.　Vine a moderate
grower, with yellow brownish wood, marked with darker spots.
Leaf medium, thin, smooth, pale green above, whitish green
below, short stem, young shoots white and wooly.　Bunch
large, with uneven sized berries; berry small, oblong, transparent, yellow, covered with gray bloom, very juicy; stem
short and stout.

In conclusion of the description of the leading white wine
grapes, I wish to say that I could have added a great many
more of high promise, many of whom will no doubt prove
valuable for certain sections of the State.　But I did not wish
to make too large a list, nor did I wish to recommend anything for general cultivation that has not been well and
thoroughly tried.　Among them I will name the famous
White Pinot (White Burgundy, Melon blanc) the Furmint or
yellow Mosler, Sauvignon blanc, Wests White Prolific,
Moselle Riessling, etc.

MUSCATELLE TYPE.

The making of this class of wine has so far not been fostered much, as the offensive rankness of flavor in the Muscat

of Alexandria, generally used in white wines of that type, has prejudiced the public against them, so that there is but little demand for them now. Aside from the Muscadelle de Bordelais described below, which, however, has an entirely distinct flavor and bouquet, there is but one variety now cultivated, which will make a very fine wine, dry or sweet, as it may be handled in cellar; this is the *White Muscateller*. Synonyms Gelber Muscateller, Muscat blanc, Muscat de Frontignan. This is mostly cultivated in Southern Europe, in Germany, Spain and France, and its wine classed very high, both in the dry and sweet form. The famous "Muscat Lunel" of France, one of the choicest sweet or liqueur wines, which retails as high as $3 per bottle, is made from this grape, and there is a certain class of customers here, who prefer its dry wine to any other, and pay high prices for it.

Vine a strong, upright grower; wood reddish brown, straight, with rather long joints. Leaf round, seldom lobed, generally heart-shaped, medium size, smooth, light green above, paler green below. Leaf stem thick and short, cords of young shoots greenish brown, slightly wooly. Bunch medium, narrow and long, compact, sometimes shouldered; short thick stems; berry medium, round, yellowish green, often acquiring a brown tinge in the sun, covered with white bloom, thick skin, very sweet, with a pronounced but delicate Muscat flavor. This may yet become a very profitable variety in this State, as it is a good bearer with long pruning, and fine wine has been made repeatedly from it. It seems to succeed well in the more southern sections, and more attention should be paid to it there, for the manufacture of liqueur wine

WHITE WINES OF SHERRY TYPE.

There are quite a number of the recent importations from Spain and Portugal, especially among those made by the

Natoma Co., which promise highly, but they have not been
sufficiently tested yet to speak definitely as to their merits.
I have already referred to the sherry flavor which the Sultana
acquires with age, and have no doubt that it could be used
for that purpose as well as for light wine. West's Prolific is
evidently of this class, although it seems difficult to find its
true name, and is a variety that will not disappoint the
grower, as it seems an unusually heavy bearer, and' makes
fine wine and brandy. I am not sufficiently familiar with ti
to give a definite description of it, but have seen very fine,
high flavored wine and brandy from it, made by Mr. West, at
Stockton, San Joaquin county.

Palomino. Synonyms, Listan, erroneously known as
Golden Chasselas in Napa Valley. This has been cultivated
here for a long time under the last name, and has acquired
quite a reputation as an abundant and regular bearer, also
making a good white wine, which, however, always acquires
more or less of the sherry flavor with age. Vine a fair
grower, wood close jointed. Leaf medium, oblong, deeply
lobed, bright green above, grayish green and tomentose
below, stem short, young points with reddish tint and wooly.
Bunch large, conical, rather loose and shouldered. Berry
round, full medium, sometimes flat, pale green with yellowish tinge, thin skin, juicy and sweet, resembling Chasselas,
which has perhaps led to the misnomer. A profitable and
reliable variety. Ripens latter part of September.

Yellow Mosler. Pedro Ximenes, erroneously. This is one of
the celebrated sherry varieties of Spain, and has been cultivated here with variable success. Mr. Crabb reports it as a
good bearer on his place, while on the Talcoa vineyards, exposed to the strong winds from the bay and coast, it suffered
badly from coulure, although the vines showed abundance of
fruit and certainly make a splendid wine, delicate, smooth,
and high flavored. It is well worth a trial in locations where

it is somewhat sheltered from wind. Vine a strong, up-right grower, with grayish, short jointed wood. Leaf large and heavy, lobed; bright green above, paler green below, covered with fine hairs or wool, young points tomentose or wooly, whitish. Bunch long and loose, somewhat shouldered, stems of berries very long, berry oblong, yellowish white, full medium, transparent, skin thin; juice sweet and aromatic, but also with a lively acid. Ripens somewhat late, about first week in October.

Among the most promising varieties of that classs, now under trial, are Mantuo de Pilas, Mourisco Blanco, etc. All this is as yet experimental, and needs further development.

RED WINE VARIETIES.

Here again, we do not suffer from scarcity of varieties, but in fact there are so many that it is very difficult to choose from them. I shall therefore confine myself mostly to those which are well proven to succeed in this State, and give a list of untried but promising varieties afterwards. Let us not for-get, however, that the high character of a red wine depends largely, if not altogether, on the soil which produces it. It is useless to plant a red wine grape on sandy soil, lacking in the minerals, the substance that will alone produce color, tannin, and also that fine flavor which a good claret or burgundy should have, to compete successfully with the best brands of Europe. I have already given the leading ideas in Chapter I, the reasons which lead me to believe that we already have an over production of *vin ordinaire*, of the common grades of red or cargo wines; let us not add more to them, but be careful where and what we plant. A high grade wine costs no more for casks and making, and it is even handled easier in the cellar, and with less labor than a common one. It costs no more freight or casks to ship it, yet it will bring double and treble the price, and what is more, will increase the demand and build up the reputation of the State and the

individual grower. And I say frankly, that I do not write
this book for those who, other things being even, would not
take more delight in handling and producing fine wine, than in
producing a common or low grade, even if equally profitable.
Such men will not add to the prosperity of the business nor
their own, and the sooner they step out of the ranks, and
make room for better men, the more lucky for us and them.

Zinfandel, or Zinfindal, as some call it. The true origin
and dissemination of this important variety is not yet clear.
It seems clear, however, that Col. Agoston Haraszthy brought
it from Hungary, and that it was also received from some
New York nurseries about the same time. Downing in his
" Fruits and Fruit Trees of America " describes it among the
the foreign varieties. Be that as it may, it has proven of great
value in developing the wine industry of the State, as it
proved that a really good, red wine, resembling choice claret,
could be made in this State, a fact which was very much
doubted before its introduction. It may be closely related to
the most famous red wine grape of Hungary, the Kadarka, the
description of which closely resembles it. However this
may be, we know and appreciate it under its present name as
one of the most valuable grapes for red wine in good locations, and properly handled. I have yet to see the red wine
of any variety, which I would prefer to the best samples of Zinfandel produced in this State. Unfortunately these *best* samples
are like angels visits, " few and far between." The reasons
for this are manifold. While it will grow and bear abundantly in almost any soil, it is by no means a perfect grape,
and must be closely studied to give its best results. First, it
needs a soil rich in minerals, iron especially, to produce its best
fruit. Then it must be well ripened, and many cannot wait
for this, but pick it when fully colored. As, with a fair percentage of sugar, it also contains abundance of tartaric acid,
it will make a wine that is greenish, harsh and sour, if

picked too early. Then it ripens unevenly, often having a
large quantity of shriveled berries together with unripe ones,
on the same bunch. This is apt to deceive the wine-maker,
as the sugar contained in the over ripe berries does not appear fully in the must, when testing with the saccharometer.
When this is the case, and there are many of these dried berries, the juice will really come to 25 Balling, when it shows but
22 to 23°. Moreover, these dried berries are a troublesome
element in fermentation, and need careful watching and frequent stirring to bring it through safe. But for all this, it is
a noble grape, and deserves all the care we can give it.

There are many locations in this State where it has been
planted, and will not make a first-class red wine, where it
could be utilized better for white wine. In this case, let
the grapes be thoroughly ripe, then crush and press immediately, but press lightly, throwing the remainder of the pomace
into the fermenting vat. together with such red wine grapes as
Mataro, Crabbs Burgundy, etc., and make them into red wine
afterwards. The first run of the juice, thus obtained, will
make a very sprightly white or rather Schiller wine (light red),
which can be advantageously used in blending with heavier
bodied white varieties. It generally has abundance of acid, and
a very agreeable flavor. But this is diverging into wine making and I will return to the subject of grapes, asking the reader's pardon for overstepping my limits here.

Vine a vigorous grower, with grayish brown wood. Leaf
dark green, lobed, with lighter green below, rather hairy or
wooly, long; leaf stem reddish, also long and wavy, young
shoots slightly tomentose, tinged with red. Bunch long and
heavy, shouldered, often double, or the shoulder as long as the
main bunch, stem short and strong, brown, compact. Berry
medium, round, black with blue bloom, and a peculiar star-
like dot in the center, but often intermingled with small shriv-

eled berries, ripening unevenly, very juicy, with a lively acid mingling with the sweet; skin thin.

The vine is very productive, easy of cultivation, often producing a second and even a third crop from the laterals. Well adapted to short stool pruning, and 3 to 4 feet stakes.

Mataro, Synonyms, Mourvedre, Catalan, Balzac, Upright Burgundy. I put this grape here, not because of its high quality, but because it forms a basis, and often a wholesome addition to many French clarets, and may become useful as a blend with Zinfandel and others, as it ferments easily, its wine is said to be very healthy, and improves with age. Otherwise it rather produces *quantity* than *quality*.

Vine a strong and very upright grower, wood brown, but with gray bloom. Leaf thick and heavy, medium size, dark green above, light green below, tomentose or woolly. Young shoots whitish gray, with many strong tendrils. Bunch large and heavy, shouldered, stem very thick and woody. Berry rather small, round, black, with blue bloom, rather dry, with abundance of tannin, but not much color. One of the most productive and easily cultivated vines, and useful in a vineyard of red varieties in many respects. Ripens late, middle of October.

Refosco, Synonym *Crabbs Black Burgundy*, Petit Pinot. This may not be its true name, it may be Pinot Noir, but it is known best by the second name. It is a very productive variety, bearing well with short pruning, and makes a very deep colored wine, which is of high character, though perhaps lacking sprightliness, which can be remedied by blending with more sprightly varieties.

Vine a moderate, but very symmetrical and upright grower, with grayish brown, short jointed wood. Leaf rather small, heart shaped, seldom lobed ; dark green above, whitish green beneath, young points whitish, tinged with red. Bunch small but compact, some times shouldered, with short stems ; cylindrical, berry small, slightly oblong, black with blue bloom, moderately juicy, sweet, rather thick skin. Ripens about first of October.

Gamay Teinturier, Synonym Gamay McGuey. This is a very productive variety, though a moderate grower, and as its juice is red, is no doubt valuable as a grape for coloring, also makes a very sprightly and finely flavored wine by itself, of the true claret type.

Vine a rather slow grower, with close jointed, dark colored wood. Leaf medium, heart shaped, shining dark green above, lighter green below. Leafstems short, reddish. Bunch small to medium, compact, cylindrical, sometimes shouldered, with short stems. Berry medium, black, oblong, juicy, sprightly and high flavored, with purple juice. Colors early, but ought to hang until first of October to develop its true quality, it is a very abundant bearer, with short pruning.

Grosse Blaue, Koelner, Grobhwarze. This variety was introduced here by Mr. John Thomann, who brought it from Switzerland; and has made quite a name as a valuable variety for blending, and on account of its deep color as well as abundance of tannin. While I do not consider it a strictly choice variety by itself, yet it makes a good neutral wine, which is very useful as a blend with Crabbs Black Burgundy, and other softer varieties, and therefore very useful in the wine cellar.

Vine a vigorous but not very stocky grower, wood grayish brown, long and thin, close jointed. Leaf thin, deeply lobed, purplish green above, whitish green or tomentose below, stem long and thin, purplish green, young shoots purplish white. Bunch large and cylindrical, sometimes shouldered ; berry, large, oblong, black, shining, of pleasant quality for the table, juicy, stem of bunch long, moderately productive. Ripens about first of October. Needs rather long pruning.

Petit Sirrah. This, although of recent introduction,

seems to succeed very well here, and fine wines have been made from it. It needs somewhat long pruning and high stakes to bring out its full bearing qualities. Vine a strong, long jointed grower, wood grayish, with brown dots. Leaf large, lobed, rough, dark green above, light green and tomentose beneath, young points greenish white, tomentose. Bunch full medium, shouldered, with rather long stem; berry oblong, medium size, black with blue bloom, skin rather thick, moderately juicy, good flavor. Ripens about the first of October.

Mondeuse, Gros Sirah. This is closely related to the foregoing, almost identical in growth and leaf, but a heavier bearer, a more compact bunch and larger berry. Said to make a somewhat coarser wine than Petit Sirah, but very valuable for blending. Ripens about the same time.

Carignane, Synonym Crignane. This variety has proved a fine grower and very abundant bearer here ; its young wines rank with the finest reds I have tasted in the State, but it is said to deteriorate with age. If this should be so, and it seems to have the same record in France, there are certainly ways and means of counteracting this, by judicious blending with other varieties. Vine a strong grower, wood yellowish brown, with white spots, young shoots green, tomentose, tinged with red. Leaf large and thick, nearly heart shaped, dark green and shining above, grayish green and tomentose beneath, leaf stem thin. Bunch very large, moderately compact, shouldered, stem long ; berry medium, slightly oblong, black with blue bloom, thick skin, but sweet and juicy. Ripens about first of October.

Cabernet Sauvignon. This is the highest type of Bordeaux claret, but unfortunately it is a shy bearer. Its aroma is so peculiar and distinct, however, and at the same time so strong, reminding of the frost grape flavor in the Clinton and Canada, that a small proportion of it in fermentation will give its peculiar character to other varieties rather deficient in flavor but good bearers. How far this can be carried, and with what varieties it would make a good blend, remains to be tried further. I would suggest the Mataro and Carignans.

Vine a slender and rather weak grower, wood brown, with a grayish cast, leaves light green, deeply lobed, rather small, downish beneath, laterals abundant and small, points of shoots gray with reddish tinge. Bunch rather small, loose, shouldered; berry small, round, black, covered with blue bloom, juicy and sweet, but with a peculiar aroma referred to above. It is subject to Coulure, and bears small crops generally, even with long pruning, but can hardly be dispensed with, on account of its high character, which it will impart to other varieties in fermentation.

Cabernet Franc is closely related to it, but the leaves are not so deeply lobed, and the grape of perhaps not quite so high a quality, though it seems somewhat more productive.

Chauche Noir. Synonyms, Blauer Burgunder, Blauer Clævner, Black Pinot, Black Cluster, Black Riesling, Pinot Noir, Black Morillion. This is one of the most famous red wine grapes of Europe, forming the basis of the most renowned French and German wines of the Burgundy type. It is not a very heavy bearer, however, nor is it very intense in color, and I believe that its true province *here*, is to make a fine white wine from its first pressing, and the pomice after the pressing of this to be added to wines of deep color, such as Zinfandel, Grosse Blaue, etc., to give bouquet and finesse.

Vine a strong grower, stocky and heavy, with many branches and laterals, close jointed; wood brownish gray, with black spots, buds close, 2 to 3 inches, grayish, woolly. Leaves medium size, roundish, with 3 to 5 lobes, dark green above, lighter green below, tips of young shoots reddish gray, tomentose. Bunch small, sometimes shouldered, compact; berry slightly oblong, black, with slight bloom, small, skin thick, moderately juicy, fine flavor. Requires age, and long pruning to produce well, and is seldom a heavy bearer, but will make very choice wine. Ripens early, and as it also starts early in spring, is susceptible to late frosts.

Meunier, Synonyms, Millers Burgundy, Muellervebe. Vine a strong grower, resembling the foregoing in the shape of the leaf, and habit, but the leaf is covered with white bloom, like flour, hence its name. The same bloom is prevalent on the berry, which makes vine and fruit look like if flour had been dusted on them. It makes a fine and delicate red wine, and is somewhat more hardy in its bloom than the foregoing, though it also needs long pruning to bring a fair crop. Ripens early, about 10th of September here.

Portugieser, Blauer. Synonymns *Moreto.* Vine a strong grower, with strong, pithy, young canes, which look almost flat, wood brown, with darker spots and streaks, young points shining light green. Leaf large, thin, deeply lobed, round, smooth, dark green above, light green below, shining. Bunch medium size to large, compact, sometimes shouldered, short, woody stem; berry round, medium, blackish blue, with fine bloom, some dark rusty spots, skin thin, very juicy and sweet, ripens early. It makes a very pleasant, dark red wine, without prominent character, and should be blended with grapes which contain more tannin and acid, or used for Port, for which it is very good. The vine is very productive.

Trousseau. Synonym, Trussiaux. Vine a strong grower, productive with long pruning. Leaf medium, nearly round, not lobed, or but slightly, downy below, deep green above. Bunch rather small, compact, seldom shouldered, short stem; berry below medium, slightly oblong, short stem, ripens early, very sweet. Better adapted to Port than for Claret. For the first, it is perhaps better than any other variety.

Tannat Noir. This new variety seems to meet with general favor, being productive, hardy, and making a very fine, dark red wine, with a good deal of tannin and character.

Vine, a good grower, productive. Leaf rather large, rough, tomentose, slightly lobed. Bunch medium, shouldered, compact. Berry oblong, blue, full, medium, ripens late.

Spanna. Synonyms Nebbiolo, Nebbiolo D'Asti. An Italian grape from which some of the most renowned wines of that country are made. Wood vigorous, light brown, long jointed; leaf large, tomentose, deeply lobed, stalk long. Bunch medium to large, long, shouldered, loose, long stem; berry roundish oblong, violet blue, thick skin, sweet and juicy, ripens late.

There are others that are very promising, but not fully tried, and I think it best not to make too long a list. The above will give choice enough, and also comprise some of the best grapes for Port. Among those for coloring especially, and worthy of a trial, I will name Pied de Perdrix, Petit Bouschet, Alicanthe Bouschet, St. Macaire, Ploussard. Among the varieties which make an excellent wine, but are so unproductive that they will not pay for planting, I will name the Malbeck, which nearly always drops its fruit with coulure.

GRAPES FOR THE MARKET AND RAISINS.

I shall take these together, as some of the best market grapes also make good raisins, and vice versa. Nor do I claim to be so well versed in them, having made the best *wine* grapes my special study. Yet the shipping of grapes to market, as well as curing them for raisins, present two very important branches of the industry, paying better at this time, perhaps, than wine-making. Besides, many have consciencious scruples against making wine, who yet would like to engage in grape growing, and for these the shipping of grapes and raisin-making offer a field which they can enter without a twinge of conscience. While I do not share their views, but believe in the introduction of pure light wines as the bes promoter of true temperance, yet I respect the honest convic-

11

tions of anyone, and feel that their interests are fully entitled
to consideration in a book dedicated to the promotion of *all*
the branches of grape culture.

The shipping of grapes to Eastern markets has received a new
impetus since we have better terms and facilities. How to pick
and pack them, and at what time, we will consider in a special
chapter, and now simply discuss the best varieties. While
many may be used for home markets which will not stand an
Eastern trip, in shipping grapes we must confine ourselves to
varieties which are attractive in appearance, and have at the
same time a rather thick, tough skin, so that they will carry
without bruising. *Quality* is a secondary consideration, for
the most high flavored and delicate grapes, if they do not
carry well, will not bring a paying price in market if they arrive in bad order. For home use, of course, there are a number of varieties of choice quality that will not bear transportation. Nearly all our choice wine grapes are also nice for
the table, often more spicy than those which are adapted for
transportation.

That the *climate* is also all important in the choice of varieties is self-evident. Some localities will find greater advantages in raising *early* varieties, others in planting the very latest.
Some of the Santa Cruz mountains shipped grapes fresh from
the vines as late as Christmas and New Years last year, having had no killing frosts; and they were sold at $2.50 per
20 ℔ box in San Francisco at wholesale, certainly a very
paying business. Vacaville and Pleasant Valley generally ship
the earliest, and obtain high prices for them. But the early
varieties and early locations will pay best for home market;
for shipping East they will come into competition with Eastern American varieties. Therefore, for shipping to the East
we should have either very showy varieties, which find a ready
sale on account of their size and beauty; or very late keepers
to follow *their* latest varieties. The southern route now fur-

nishes facilities for shipping nearly all winter; and thus our late varieties could come into the Eastern markets for the winter holidays, and would outsell any of their late varieties.

So much for general rules to guide in selecting varieties for this purpose, now to the discussion of the varieties themselves.

FOR EARLY MARKET.

The Chasselas Fontainebleau, or Sweet water, is still about the earliest variety for that purpose, which bears sufficiently. This has been described under varieties for wine. The *Early Madeliene*, synonym Madeliene Angevine, is still a week earlier, and of rather better quality, will also carry well ; but so far it has been a shy bearer, especially where late frosts prevail, as it is one of the first to vegetate and bloom. For locations free from frost, it would pay to try with long pruning, when I think it will bear well. I would certainly try it at Vacaville and Pleasant Valley. Vine a moderate grower, with long jointed, brown wood. Leaf medium, deeply lobed, dark green above, tomentose below ; young points reddish, woolly, slender. Bunch medium, compact, shouldered. Berry medium, oblong, yellowish green, transparent, rather thick skin, sweet and juicy.

The earliest black variety is perhaps the *Black Malvasia*, so called in this State. It is a much better table and shipping grape than it is a wine grape, and as it is very productive as well as showy, will pay to grow for market to a limited extent. It is too well known to need description here, in fact has been too largely planted for wine, for which it is not adapted, as especially the red wine will deteriorate with age, and can only be used for port. If pressed lightly, as soon as crushed, the first juice makes a very pleasant light, white wine.

FOR LATE MARKET.

The most prominent among these is the Flame Tokay, which, on account of its handsome color, magnificent berry and bunch and good shipping qualities, is perhaps the most profitable where it succeeds and colors well. This is not everywhere the case, however; it wants a warm and rather moist soil, cold locations will not do for it, and yet in the most southern locations, it is also apt to sunscald. This can, in a great measure, be prevented by early summer pruning, and I shall refer to the subject under that head.

It was first introduced into the State by Wm. McPherson Hill, of Sonoma Co., and I cannot find it in any work on Grape Culture I have examined. Vine a strong grower, large in all its proportions, wood, joints and leaves. Wood dark brown, straight, with long joints; leaves dark green, with a brownish tinge, slightly lobed; bunch very large, sometimes weighing eight to nine pounds, moderately compact, shouldered; berry very large, oblong, red, covered with fine lilac bloom, fleshy and crackling, firm; ripens late. The clippings of small and imperfect berries, cut out when packing, will make a very agreeable white wine, with pleasant acid and good bouquet, and can thus be utilized to good advantage.

Black Damascus. Synonym Blauer Damascener Zwetschgentraube, Ribier. Vine a medium grower, wood light brown, striped with darker brown, short jointed. Leaf round, five lobed, smooth, light green above, tomentose beneath, stem reddish; bunch large, loose, shouldered; stem large and long, woody; berry very large, oblong, dark blue, covered with lighter bloom, meaty, skin thick, ripens late.

Emperor. Vine a strong grower and rather a shy bearer, better adapted to Southern than Northern culture, as it ripens very late. Wood long jointed, brown, half rough and large, deeply lobed. Bunch long and loose, shouldered, very large; berry oblong, purplish black, covered with lighter bloom, thick skin, firm.

Black Cornuchon. Synonym, Cornichon Violett, Eicheltraube. Vine a strong but stocky grower, with thick, close jointed, brown wood. Leaf large and thick, deeply lobed, dark green above, grayish green and tomentose below, five lobed, young shoots light green, with tomentose points. Bunch very large, loose, shouldered, with long stems and drooping ; berry large, long, dark blue, with lighter dots, fleshy, thick skin, very late.

Rose of Peru. This is a very handsome and productive grape, of good quality, but does not carry quite as well as some other, vine a strong grower, resembling Mission, with dark brown, short jointed wood; leaf deep green above, lighter green and tomentose below. Bunch very large, shouldered, rather loose; berry round, large, black, with firm and crackling flesh, ripens rather late.

Gros Colman. Synonym Dodrelaba. Vine a very strong grower, long joints, dark brown wood. Leaf very large and thick, more broad than long, slightly lobed, dark green above, white and woolly below, young shoots tomentose. Bunch heavy, broadly shouldered, rather loose ; berry very large, round, black with blue bloom. Ripens very late but evenly, and is very productive, but may not carry so well as some others.

Black Morocco. This is used as a shipping grape further South, not adapted to the North, where it ripens very late and unevenly. Vine a straggling, drooping grower, with numerous laterals, which generally bring an abundance of second crop. leaf thin, deeply lobed, and serrated, dark green and shining, Bunch very large, rather compact, heavily shouldered; berry very large, black, fleshy, of rather poor quality.

Muscat of Alexandria. Synonym Moscatel Gordo Blanco, weiser Muscat, Damascener, etc. It is yet a disputed point in this State, whether Muscat of Alexandria or Muscatel Gordo Blanco are the same. In the books on grape culture they are called identical, and I can see no difference here. While this is the leading raisin grape, here and abroad, it is also a very important shipping grape, as it carries well, looks well, and many admire its peculiar flavor. It can also be utilized to make the well known sweet Angelica wine, or rather Cordial, for which the clippings of small berries could be used when packing and making raisins. Vine a short, rather straggling and bushy grower, well adapted to short stool pruning, as it forms rather a bush than a vine, wood gray, with darker spots, short jointed. Leaf round, five lobed, bright green above, lighter green below, young shoots a bright green. The laterals produce a second and even a third crop, and the second crop will often ripen to be fit for shipping. Bunch long and loose, shouldered; berry oblong, light yellow when fully mature, transparent, covered with white bloom, fleshy, with thick skin, very sweet, and a decidedly musky flavor liked by many, and disliked by others.

Malaga, Synonym, Weisner Damascener, frueher weisser Damascener. Vine a strong grower, wood reddish brown, short jointed. Leaf medium, leathery, smooth, deeply lobed, light shining green above. Bunch very large, loose, shouldered, long, stem long and flexible; berry very large, oval, yellowish green, covered with white bloom, thick skin, fleshy. Ripens rather early and also makes good raisins.

Verdal. Synonym Cheres, Malvoisie de Sitjo. Vine a strong grower, long joints. Leaf large, deeply lobed, tomentose; bunch short, heavily shouldered. Berry oblong, yellowish green covered with fine bloom, ripens late, very productive.

White Cornichon. Synonym Cornichon blanc, weisse Eicheltraube. Vine a strong but short jointed, stocky grower, light brown with darker buds. Leaf long, thin, deeply lobed, light green above, tomentose below. Bunch very large, loose, with long drooping shoulders. Berry oblong, golden yellow, with light dots, thick skin, fleshy and transparent, ripens late.

The Sultana has been described among the wine grapes. It makes very fine seedless raisins, but they do not seem to sell as well as the Muscatel, very likely on account of their

small size. The white and black Corinth, from which the
Zante Currants are made, do no seem reliable here, and so
far have not proved profitable.

--- -- -

CHAPTER XIII.

CULTIVATION AND TREATMENT DURING THE FIRST AND SECOND SUMMER.

After the vineyard has been planted, it should be kept well
cultivated, the surface kept loose and mellow by frequent stirring with plow, cultivator and harrow; it is the only method
by which moisture can be kept up, and the vines can live
and grow freely. If the soil has been well and deeply prepared,
it will need no deep plowing the first summer, unless late
rains have hardened it down after planting. It is generally
sufficient in all ordinary soils to run a two horse sulky cultivator, in which the shares are so arranged that the two middle ones take one side of the row each, while the horses also
walk one on each side of the row, (or straddle it, as the common expression is). If the operator is careful, he can come
very close to the vines, and by setting the shares or shovels
so that they will throw the ground slightly towards the vines,
they will get loose earth around them. Should the ground
have become hardened, one of the numerous vineyard gang
plows will have to be run through the row, or if this is not
available, a one horse plow can be used, though this is much
slower work. One of the most convenient gang plows for
vineyard work is (Fig. 12,) manufactured by H. Hortop, Ruth-

Fig. 12.

erford, Napa Co., who is a good mechanic, and living in the midst of the main wine-growing district, has had a chance to study the wants of the wine growers. It has two shares, and in fair soil can be drawn by one good pair of horses. The shares used, Oliver No. 8, can be taken off and used for single plow. The plow is calculated to finish four furrows in a round, going up on one side of the row, throwing the ground to the middle, and returning on the other side in plowing from the vines. The wheel in front regulates the depth. By plowing one furrow with single plow close to the vines the row is finished. In plowing to the vines the single plow, or one horse plow, is used first, throwing the ground to the vines, and one round with the gang plow, going up on the right hand side and returning on the same row will finish it.

Fig. 13 is a plow for deep tillage, made by the same party, for breaking ground for vineyards. It is very strong, will stand the draught of six horses and mules, and calculated to run sixteen to eighteen inches deep. Where the soil is not too tenacious, this will be found to save time and labor.

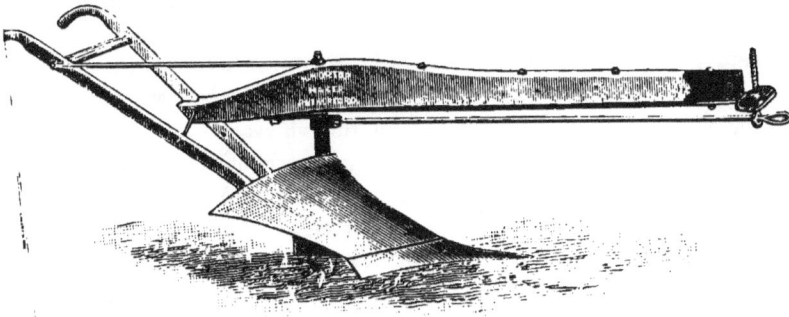

FIG. 13.

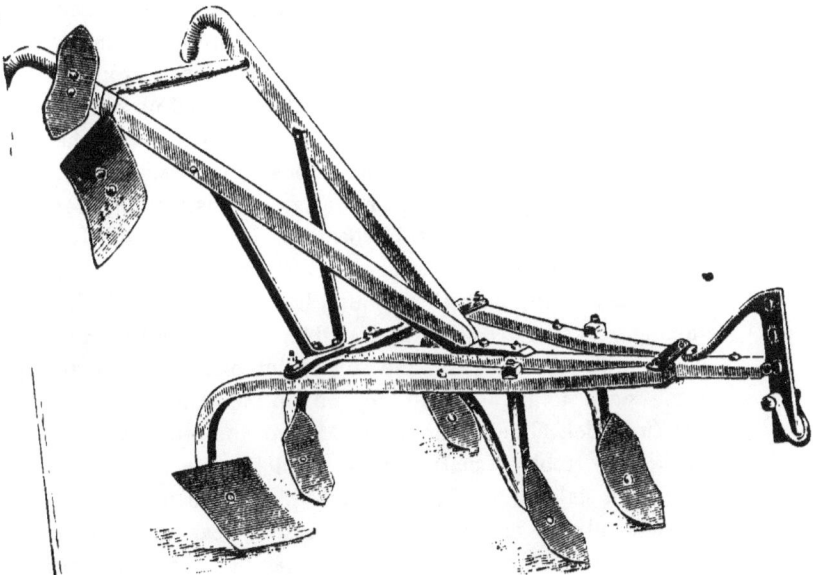

FIG. 14.

(Fig. 14).   Cultivator for two horses, to follow the plow in later cultivation, also made by the same party.   The shares with small mouldboards at the sides work like small plows, throwing the ground to the vines, can be changed ad libitum for the longer pointed shares, also for a weed cutter blade, to be attached to the shanks behind, to extend all the way across, with three cultivator shares in front of it to loosen the soil.   This is very useful for late cultivation, especially where the morning glory, that pest of California vineyards, prevails.

Either the harrow or the clod crusher should follow the plow or cultivator if the ground is at all lumpy, to break the clods and make a mellow and even surface.   The later this is kept up in summer the better will be the growth on the vines; they will grow all the better if cultivated *all* summer, and they will pay well for liberal treatment by early and abundant crops.

If resistant vines have been planted, to be grafted when strong enough, they will need no pruning the following winter, as it will be just as well, and make them stockier and stronger, to let them grow unchecked.   They will also need no stakes, as vines which are not irrigated, make but a small growth the first summer, until the roots have become firmly established. If viniferas have been planted, however, or resistant vines for direct production, such as Herbemont, Rulander, and Lenoir, the young vine may be cut back to two buds of its last summer's growth, and should also have a stake for future training.   Where redwood is available, that is the best and cheapest, as it works easy, and is very durable.   The length of the stake to be used will depend on the variety planted. For Zinfandel, Green Hungarian, Mataro and other varieties adapted to stool or short pruning, (which some call goblet pruning) a stake of four feet is amply high enough, and this can generally be removed the fifth or sixth year, as the vine will support itself then.   For varieties requiring long pruning five and even six foot stakes are necessary, and will be found most economical in the end, as the vines will pay for the additional cost of the stakes in their first seasons bearing. The comparative price of the stakes is *here* about as follows, at the lumber-yards :

4 foot stakes, sawed, 1½ inch diameter, per 1000. .... $16 00
5   "   "   "   1½ "   "   "   " 23 00
6   "   "   "   2 "   "   "   " 34 00

If we also count in the additional cost in hauling and handling, we will have about 2½ cents more, for 6 feet than 4 feet stakes. Two pounds of choice grapes additional the first season, will pay for this, and square the account, with the additional product every season, which at a low calculation will be 5℔ per annum in favor of the longer stakes.

Drive the stakes on the side of the vines from which you have your prevailing winds in summer. Of course they should be pointed, and can then be driven with a sledge hammer or wooden maul, when the ground is soft in winter. They should be in the ground from 15 to 18 inches, so that a six foot stake would stand about 4½ feet above the ground. If the vines have made but little growth, the staking may be postponed until the second year, and the young vine tied to the marker the second summer, The best material for tieing in summer are the leaves of the Phormium tenax, or New Zealand flax, which can be torn into strips, and are very flexible and strong. The leaves of the common Dracena (Dracena Draco) or Dragon tree, answer about equally well. In fact, all the Dracenas and Yuccas furnish excellent tying material, better than the common grape twine now in use, and not near so expensive. Every vineyardist should plant some of these, especially of the Phormium, which is gratuitously distributed by our State University. They are all fine ornamental plants, and only the dead or dying leaves need be used.

The second summer, our first operation must be the plowing. Some of the vineyard gang plows, described before, are available, they can be used for the work in the centre of the rows, setting the share on each side so that they will throw the ground together in the centre, and away from the vines. If only a common two horse plow can be used, commence by plowing a furrow exactly in the centre of the row, then in returning, throw the next furrow against it, and from the row. Go down on the other side, throwing the furrow on the first, and away from the row. Then finish up with a one horse plow, in the same manner, as close as you can come without injuring the vine, letting the horse walk in the furrow plowed before. This will leave but a narrow strip of, say six inches, which can easily be finished with hoe or spade. The two pronged German hoes or karsts are a good implement for that purpose. Even where the gang plow has been used, it will always be well to finish with the one horse plow, as it admits of closer and more careful work. The plowing in the centre can vary from 4 to 6 inches in depth, according to the nature of the soil. Close to the vines, 3 or 4 inches will be sufficient ; and the hoeing or spading should not be deeper than to break the hard crust around the vine, caused by the winter rains, not deep enough to injure the roots.

Follow the plow with a vineyard harrow, of which a six foot Acme pulverizer is one of the best, or a revolving harrow, which will smooth and pulverize, and destroy the weeds at the same time. These operations should be performed as soon as it is dry enough in early spring, so that the ground will work well. We often have a spell of good weather in winter, which may be taken advantage of. Always try to be ahead with your work, drive *it*, when you can, so that it will have no chance to drive *you*. Plowing in winter, however, should always be done parallel with the hillside, if on sloping ground, as, if plowed up and down hill, the spring rains are apt to make deep washes, where they have the furrows as so many gutters.

When this plowing, hoeing and harrowing is done, we can cross plow, that is, if the first plowing has been done from East to West, we now plow North and South. Here we reverse the operation, take the one horse plow first and throw the ground against the vines; taking care, however, not to cover them. Then we follow with the two horse plow, or gang plow, and finish out the middle, so that every inch of soil in the vineyard is stirred when this is done. This will put the soil in good, mellow condition, and the finishing is given by taking the clod crusher or drag crossway over the rows, which breaks all lumps, and fills up the middle furrow somewhat, leaving our vines, when these operations are completed, in a bed of mellow earth, where they can grow and flourish. The young weeds and vegetation turned under serve as fertilizers, and at the same time have a tendency to loosen the soil. Cultivating from time to time, as the weeds may begin to grow, and the soil requires stirring, as well as cutting down the weeds which may grow around the vines with a light hoe where the cultivator cannot reach them, will finish the cultivation for the summer. I reiterate again, if you have the time, cultivate freely; you can not overdo this. Your vines will well repay any extra trouble you may take with them, by additional growth and fruitfulness.

CHAPTER XIV.

CULTIVATION, PRUNING, AND TRAINING THE THIRD AND
FOURTH YEAR.

The cultivation will be essentially the same, although, as
the vines increase in size, we cannot come so close to them
with the plow, and therefore must hoe somewhat more.   But
as pruning and tying in larger vines must be done before cultivating, we will consider this principally, following up with
summer pruning or pinching, and tying the young growth in
summer.

STOOL OR SPUR PRUNING (goblet pruning).

We have many varieties for which this is a very convenient
and easy way of training, and which will with this mode of
treatment, the simplest of all, produce abundant crops.

VARIETIES ADAPTED TO THIS TRAINING.

*Marsanne, Green Hungarian, Clairette Blanche, Victoria
Chasselas, Burger, Folle Blanche, Palomino, Zinfandel, Mataro, Refosco, Gamay Teinturier, Blauer Portugieser, Flame
Tokay, Gros Colman, Muscat of Alexandria, White Cornuchon,* and of the *old* varieties, *Mission* and *Malvasia*.   Of
course, there may be many others, but my experience does
not warrant me in recommending them for this treatment.
Four foot stakes will be sufficient for them, and they can be
put in in most cases when the vines are pruned.

The proper *time* for pruning is when the leaves have dropped
in fall, and all during the winter months, as our winters are
not severe enough to damage the wood or buds.   The sooner
we commence with it the sooner our vineyards will be ready
for the plow.   Some defer it until late, as a preventative

against late frosts, but I hope to show my readers a better method under that heading.

We will suppose our young vine to have at least one good stocky cane of three to five feet long. This we cut back to two feet, about half an inch above a bud, making a slanting or vertical cut above the bud. (Fig. 15) shows the vine,

FIG. 15.   FIG. 16.

the cross line indicating where to prune. (Fig. 16) shows it pruned, staked and tied. For the upper tying I have found annealed wire No. 16 the best, most convenient, and cheapest material. It can be used several years, is easily applied, quicker than any other, as the wire is cut to the desired length with a cold chisel, tied in convenient bundles, and the two ends are simply hooked into each other by a quick twist with the fingers. It is just as easily taken off, pulled straight, and used another year; and costs ten and one-half cents per pound. Care must be taken not to apply it too tight, so as to allow for the growth and expansion of the vine, as it will cut into the wood and bark if drawn tight. This will not break through the chafing of the vines in strong winds against the stakes, and is much cheaper than grape twine and better. Always tie just below the upper bud, so as to keep the vine firmly to its place. If tied lower, the growth above the tie will cause the vine to lose its balance and lop over, thereby making an ugly bend, where it is apt to break off. For

the lower tie either the Phormium tenax, Drcena leaves, or
the golden willow, (Salix Aurea), can be used, of which each
grape grower ought to make a plantation along the ravines
and gullies, where they will not take up any room, and can
be cut every year; the young twigs are very soft and pliable.
The silver leaved willow, growing wild on many of our streams,
is equally tough and serviceable. For want of cheaper and
better material, use grape twine. Wire is not advisable for the
lower tying, as it will cut into the vines. The young resistant
vines, if they have attained a diameter of from half an inch
to three quarters, should also be grafted, as described before,
in the chapter on grafting. April and May will be found the
best time to do this.

When the vines have been pruned and tied as indicated
above, they will bring their strongest shoots or branches from
the upper buds. Of these, two or three of the strongest
should be left, to form the future head or stool ; if the vine is
rather weak, leave but two ; if stronger, three, all the rest
which may appear from the lower buds, should be rubbed off.
It will be found advantageous to pinch off the tops of the remaining, when they have grown about a foot ; they will then
throw out laterals, and become more stocky and bushy; especially is this advisable in windy locations, as the winds have
less power on many shorter shoots, than on a few long ones.
Besides it shades the vine and the fruit, and prevents sun-
scalds ; as the laterals always come from the axils of the
leaves opposite the bunch and thus shade it.   But do not follow *late* summer pruning, lopping off the woody shoots with
knife or sicle, which is unfortunately practiced too much. It
is barbarous to the vine, causing the cane to die back, and
the fruit to ripen unevenly. The sooner summer pruning can
be done, the more beneficial it is to the vine ; and besides, it
is done so much more rapidly and easy. In half a minute a
man (or woman either) can go over a large vine in May or

June, at the same time rubbing out all barren and superfluous shoots (suckering). When the foliage is once fully developed, it is much more difficult to look through the vine and do the proper thinning, besides the first will develop so much more evenly and perfectly. In fact, winter and spring pruning are but the beginning of the training ; if not followed by judicious summer pruning and thinning, it is incomplete. The *fourth* winter or spring we find our vine with two or three strong shoots from the upper buds, presenting a miniature tree or bush. These we cut back again to three buds each as shown in Fig 17 at the cross lines, and tied to the stake, as in Fig. 18.

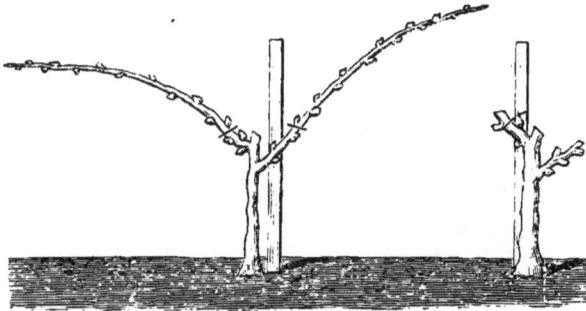

FIG. 17.   FIG. 18.

We call these spurs, and from each of these spurs we expect at least two strong shoots, from the two upper buds on each. As the buds on the vine grow in triplets, the main or fruit bud in the centre, with two smaller buds, one on each side, it is often the case that two of them or even all three will start and grow. Only one, and this the strongest, should be left from each bud, all others rubbed off, nor should more than two be left to each spur, so that we double the amount of shoots or canes we had the second summer. Summer prune as the summer before, first thinning out the superfluous shoots, by rubbing or pulling them out at their base, and then pinch the remainder.

The fifth year, we double again, as shown in Fig. 19, which
will give us four to six spurs, according
to the strength of the vine, which, for
very heavy bearers and in ordinary soil,
will be about enough.   If the vine is very
vigorous, and the soil strong, however,
we can keep on increasing the number of
spurs even to twelve ; this must be left to
the discretion of the intelligent vintner,

Fig. 19.

as it is impossible to give any fixed rule as to number of buds
to be left.   In fact, the health and vigor of the vine depend
largely on pruning according to individual strength.   As long
as a vine makes a vigorous, well ripened growth of wood,
ripens its fruit evenly and well, developing the full amount of
sugar, it has not been overloaded.   But when the growth
decreases, the berries and bunches become smaller, and ripen
unevenly, it has been overtaxed, and should be pruned
shorter.   If, on the contrary, the vegetation is too rank, the
berries abnormally large, it shows that it was not pruned long
enough, and it will suffer easier from coulure and mildew, its
wood will not ripen so well, nor be so fruitful.   The results
of the vintners labors depend largely on his nice discrimination in pruning and summer pruning.   A man who is not, to
a large extent, able to judge the capacity of a vine when he
looks at it, is not fit to prune it, and will do more harm than
good in a vineyard.   My rule is, to prune full as long as I
think the vine is able to bear, should it show more fruit than
I think it is well able to bear, I thin with an unsparing hand
in summer pruning.   It is always easier to rub off a superfluous shoot or bunch, than to add one when they are "not
there."   But do not let greed, or the desire of an immense
crop, stay your hand, when you know that your vines have
too much.   Thin out evenly, or your crop will be poor next
year and the following, vines will feel and resent such abuse.

If you want a sound man, able to do his faithful days' work every day when mature, do not overtax the boy, willing as he may be. Just so with the vine. This will apply to *all* modes of training and culture,

When we have fully developed our vine, say the sixth year, and think it has as many spurs as it is able to bear well, we prune back, that is, where two shoots have grown on a spur, equally strong or nearly so, we cut out the upper just above the lower, pruning this to three buds again, and thus we obtain the same number of spurs as the year before. The reason why we cut out the upper is, that we want to avoid elongation, but keep the vine at nearly the same dimensions, which we could not do, if we left the upper, and cut away the lower. As far as possible always prune to an outside bud, i. e. one pointing from the center of the vine, as we want to keep the head as open as possible. The rank shoots from the old wood, watersprouts or suckers as they are generally called, should be all removed in summer pruning, unless they may be needed to take the place of a failing spur or arm. This completes stool or goblet pruning. After the fourth year, the vines are generally long enough to support themselves, and the stakes may be dispensed with, to be used somewhere else.

MEDIUM OR "HALF LONG" PRUNING.

We have many varieties not well adapted to stool or spur training, which will produce well with a medium course. Some varieties are so constituted, that they will not fruit well from the first two or three buds at the base of the cane, while they will produce abundantly from the fourth to tenth bud, and some of our most valuable varieties belong to that class.

Varieties adapted for this treatment. Pedro Ximenes, (generally known in this State as Sauvignon Vert or Colombar) Chaceda Gris, Simillion, Traminer, Muscadelle de Bordelais, Chasselas Fontainebleau (Gutedel) Chasselas Violette, White Elben, Blaue Elben, White Muscateller, Grosse Blaue, Mon-

deuse, Carignan, Meunier, Tannat, Pied de Perdrix, Petit
Bouschet, Rulander (American).

For these I would recommend five foot stakes, We commence the third year, by leaving one cane, a foot to eighteen
inches long, (15 will be about a medium), pruning and tying
precisely as for stool pruning. We leave three canes or shoots
to grow from the three upper buds, which will give us something like a goblet shape to commence with, and summer
prune as in stool pruning, not quite so short however, leaving
at least eighteen inches of the young growth. Near winter
or spring we cut the three canes obtained to about 15 to 18
inches, and tie to the stake as in (Fig. 20), making the upper
tie with wire, the lower, which is only drawn around to keep
the canes from spreading, with Phormium, Dracena leaves or
or twine. The shoots from these are pinched as in stool
pruning, but one shoot from the base of each cane, left unpinched, to develop fully ; as from it we expect our cane for
next season's bearing. These three canes will each bear some
fruit, if strong enough, but should not be overloaded. Here

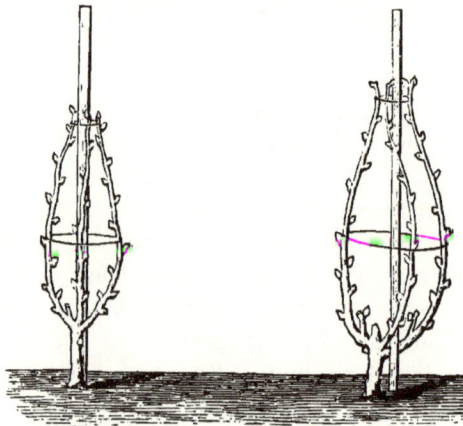

FIG. 20.   FIG. 21.

again the discrimination of the vintner is needed, and proper thinning, if necessary, must be resorted to.

The fourth winter, the old canes which have borne fruit last summer, are cut out, and replaced by the young canes from the base which were left unpinched, always leaving a spur, however, of say two buds, from the lowest branch of the old cane. (Fig 21) shows the vine pruned and tied. As the vine has become stronger, the canes can be left longer, say too feet, provided the vine can bear it, and summer pruning followed as before, with the alteration that we leave the canes for next season's bearing on the spurs at the base of the canes. This system is followed up, and is simply a renewal training, the cane from the spur taking the place of the cane or arm which has borne last season's fruit.

A modification of this treatment is sometimes followed with very good results, and makes the vine self supporting. Four canes are grown instead of three, and bent together at the top, so as to make a globe or balloon. A wire is firmly tied around them, and if the canes are equal in strength and equally loaded, the fruit hangs mostly in the middle of the globe. This method, (Fig. 22) has some advantages and some disadvantages. The circular form in which the canes are bent, distributes the sap more evenly, while with the other method, it runs more into the upper buds on canes and spurs. Its disadvantages are, that it takes more room in the vineyard, does not allow as close working, and unless the canes are of very even size, they will not balance well, when heavy with fruit, but pull to one side. We want economy in our work, especially in cultivation, and this interferes with it somewhat.

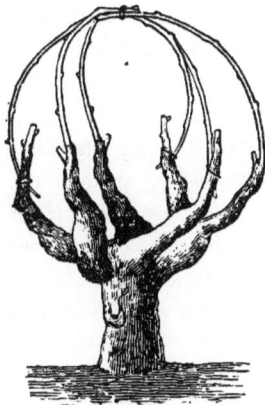

FIG. 22

LONG PRUNING AND TRAINING.

We have some varieties which are shy bearers even with this last mode of training, and which require still longer pruning to produce paying crops, yet are so valuable that we cannot well dispense with them. For these, I recommend another variation, which I have practiced with splendid results, especially on Æstivalis varieties, Herbemont, Lenoir, etc.

Varieties adapted to this treatment. Sultana, Riesling, Franken Riesling, Yellow Mosler, Petit Syrah, Cabernet Sauvignon, Cabernet Franc, Chauche Noir, Trusseau, Emperor, Herbemont, Lenoir, Nortons. It is simply a modification of medium pruning, as described before. The vine is treated the same way the third year, but for this method, six foot stakes are needed, and the three canes, started at about 12 to 14 inches from the ground, are left somewhat longer. The fourth season, instead of cutting out the bearing canes of last year, I leave these for permanent arms, to last as long as they are healthy and sound. I cut all the strong, vigorous shoots they may have, which have fully developed fruit buds, to spurs of three to four buds each, up to two feet and a half of the crown or head of the vine. This will give say three to three and a half feet. From here, I have a short cane on each arm, to reach to the top of the stake, and tie firmly with wire, with a strong tie of Phormium or twine around the middle, to hold them to the stake, and prevent their spreading. The next pruning, I leave the old arms, and from each of the spurs I select the strongest, as near the base as possible, pruning it to three or four buds; so that for each spur of the summer before, I have another, cutting out the balance. The young cane at the end of each arm, I either replace with another, or leave it, and cut its laterals also to spurs.

All varieties I have handled have produced satisfactory crops under this treatment, except the Malbeck; which always suffered from coulure to such an extent that I do not think it will pay with any manner of training, and ought to be discarded. Nor do I think we need it, with all the fine varieties now at our command. There are, of course, many other modes of training in vogue in France, Germany, and all Europe, as well as the trellis method adopted in the Eastern States, which I have followed and advocated there for many years. But the trouble with most of them is, that they offer serious obstacles to cultivating both ways, and as labor is high, we must do all we can with plow and cultivator, which not alone saves manual labor, but offers better cultivation than we can perform by hand. No hoe or spade will so thoroughly pulverize and mellow the soil as the plow, clod crusher, and harrow, where they can be used both ways. For this reason I am slow in following or recommending any method of training which will only allow cultivation one way; and the advantages it offers must be great indeed to induce me to adopt it.

THE CHAINTRE SYSTEM.

This is one of the systems which would prevent cultivation both ways, but is much recommended by French authorities, as very much increasing the product per acre, and necessitating only about three hundred and twenty-five vines to the acre, instead of three thousand, as they are generally planted there. It was Denis Lusseaudeau, at Chissay, France, who first invented and tried it, and it is enshusiastically spoken of and explained in a pamphlet with numerous illustrations by A. Vias, which was translated into English for the State Board of viticulture, and published among their transactions. Any one who wishes full information about it can obtain it from the Secretary, Mr. Clarence J. Wetmore, who has tried it himself, and thinks it well adapted to such varieties as are shy bearers and much subject to coulure. The vines are pruned in a peculiar manner, and bent over as the name im-

plies, in a trailing chain, pruning very long, as the rows are
planted in France twenty feet apart, and the vines six and a
half feet in the rows. For cultivation the whole vine is turned
around, laid over on the row, and when the ground has been
plowed the vines are turned back again, and supported by
small stakes over the empty space, so that the young growth
is a foot from the ground. They also claim better and more
even ripening, as well as a greater amount of sugar, for this
method. Mr. Wetmore would only recommend it for shy
bearers, and especially in windy locations. Others who have
tried it do not think so favorably of it. The pamphlet issued
with the second annual report will explain the method fully.
I have not tried it for the reasons given above.

CHAPTER XV.

DISEASES OF THE VINE AND THEIR REMEDIES.

Fortunately, the vine is subject to but very few diseases
here, as that terrible scourge of the Eastern and European
vineyards, the Peronospora Viticola, or Mildew *par excellence,*
has not troubled us here, and I do not think it ever will, as
long as our summers remain as dry, and our atmosphere as
pure as they are now. The Peronospora generally makes its
appearance in the East after continued rains, and murky, sul-
try weather, and then often destroys two-thirds of a crop in a
few days, an atmospheric condition which I have not as yet
observed here. The Oidium Tuckeri, or powdery mildew,
generally appears after heavy fogs, followed by dry, still weather during the middle of the day, and yields readily to the ap-

plication of sulphur, which does not stop the Peronospora, for which the remedy is sulphated copper and slacked lime. Let us hope that it may never trouble us, and pass it by.

Our Chief Viticultural Officer, Mr. John H. Wheeler, has recently published a very practical treatise on the "Oidium Tuckeri and the use of Sulphur;" which covers the ground so completely, that I take the liberty of republishing it here almost entire, and am sure that those of my readers who will study and follow it closely, will have little to fear from the disease. I have generally found one application sufficient, either at or shortly after the bloom, but the vines should be watched later on, even until the berries color, and if the spotted leaves and the grayish color which the fruit assumes, are observed, a second application is necessary. To the varieties subject to it outside of those mentioned by Mr. Wheeler, I will add the Mataro, and Marsanne. Generally speaking, all very rank growers are more subject to it than the medium or slow growers, while the Æstivalis class is entirely exempt from its attacks. Sulphuring is also a partial remedy against *coulure*, or dropping of the bloom or berry, (imperfect setting or fructification) of which I shall treat further on.

OIDIUM (TUCKERI) AND THE USE OF SULPHUR.

OIDIUM.*

The oidium *(tuckeri)* is a vegetable parasite of American origin. It attacks all growing portions of the vine and imparts to the leaves a chapped appearance, and gives them a whitish or gray color. The vine, when badly affected, has a blighted and sickly appearance ; the young leaves and tender parts be-

---

*Frequently and improperly confused with mildow, which it is not. The true *mildiou* is the dreaded *Perono-pora Viticola*, a parasite far more formidable than the disease we commonly treat with sulphur and one which does not succumb to this or other simple remedies.

The misnomer of the parasite common to California frequently leads to a confusion of remedies—sulphate of copper and slacked lime is the remedy for *mildiou*—but one not necessary for our oidium.

come dried and roll up, attacked, the herbaceous parts blacken, cease to grow, and end by withering and drying up. This latter extreme is rarely attained in California. The growing berries are attacked as readily as other parts, giving the whole a languishing and unhealthy aspect. The young branches also present blotches of a powdered nature which ultimately cover the greater part of the surface exposed to the sun, and where badly affected also taking on a whitish, powdered and eventually chapped appearance, which causes them to crack open and cease to grow. Thus it will be seen that the oidium, unlike other fungus, affects the crop directly as well as indirectly through damage to the foliage. The parasite first appears abundant in June, though frequently commencing its attack in May, at or after the time of flowering.

The conditions favoring the oidium are moisture and warmth, the latter playing the most important part. The moisture here meant is not the extremely humid condition of the atmosphere which appears with or immediately follows a rain or heavy fog continue, a condition often incorrectly named as favorable to oidium, but merely the moisture to be found in sea breeze after it has traveled ten, twenty or even thirty miles inland. An atmosphere which produces a light dew at night is sufficiently moist to favor to the utmost the propagation of oidium. Quite different in this respect is the *peronospora* and *Anthracnose* which require the deposition of heavy rain, fog or excessive moisture to produce their growth. For this reason, I believe, California has been comparatively free from the true *Mildiou*, a disease which of late years in France, where summer rains are frequent, has threatened the vineyards to as great an extent as has the dreaded phylloxera.

Our principal vegetable parasite thus far has been the oidium, one especially favored by our dry, warm climate, and one easily destroyed by the timely application of sulphur.

As before remarked, excessive moisture is unfavorable

the propagation of oidium, and a good shower will do much to remove and destroy the germs.

As to temperature, the disease begins its development where the average of day and night runs up to 53° F.; it spreads rapidly at 70° F., and is checked in its growth where the thermometer indicates near 100° F. Above 100° its damage is rapidly diminished, and at 112°—a temperature quite common throughout the interior vineyard districts of California—the germs loose their vitality and the effects of the disease entirely cease.

To be sure, where vines make a dense growth and are trained high above the ground, the germs may be so sheltered in shady spots as to escape the effects of the heat. Where, on the contrary, the vines are trained along or close to the ground and receive the reverberated in addition to the direct heat of the sun, the manner in which vines should be trained —then, the high temperature above named accomplishes a complete extermination of the parasite, a result which has often been noted in Algiers where such temperatures are frequent throughout the early growing period of the vine.

Let it be borne in mind generally that the propagation of the oidium and other vegetable parasites of the vine are greatly favored by trellises and high training. Short pruned vines and those trained close to the ground are most exempt from fungoid diseases.

A hot north wind will sweep the oidium from a vineyard well exposed to its effects. This forms one of nature's most common remedies in California, and should be a consolation to those who may otherwise lose by it in the grain field.

Some varieties of vines are found more susceptible to the attack of oidium than others, other conditions being similar

This fact should influence the vineyardist as to the frequency of applying the remedy and the amount of sulphur employed.

Varieties particularly subject to the effects of oidium are the (Muscat, Chasselas, Zinfandel, Folle Blanche, Crabb's Black Burgundy, Teinturier, Gamay, Cabernet Sauvignon, Cabernet Franc, Riesling, Carignane, Terret and Cinsaut.) Among those little susceptible are the Grenache, all of the true Pinots, the Alicante Bouschet, Petit Bouschet, Colombar, Sauvignon blanc and the Aramon. The American grapes *Vitis Labrusca*, *V. Riparia* and *V. Rupestris* are but little affected by oidium.

REMEDIES.

Many substances have been applied principally in the form of powders—lime has been extensively employed, and it has been found that any dust effects beneficial results on the diseased plant. None have proved so efficacious, however, as sulphur dust, and on this we can rest our perfect reliance, for, if properly applied, it affects all that may be desired in the way of a cure, and is comparatively inexpensive. The oidium is a disease quite easy to treat, because its spores and growth are confined to the exterior and exposed portion of the plant, which is not the case with the *peronospera viticola* and some other vegetable parasites.

THE APPLICATION OF SULPHUR AS A REMEDY.

There have been many conflicting and erroneous statements made concerning this remedy, its application and effects as applied in California. Imperfect and hasty generalizations, drawn from limited local experiences, have not unfrequently been published and results both expensive and wasteful have often followed. To correct the wrong impressions thus formed and save further dispute, it seems necessary to treat the subject in a somewhat technical manner, the truth on some points of which it seems to me precludes the possibility of further dispute as to kinds which should be employed ; the difference in the effect of various brands, imported or domestic, and the manner, and time best for making the treatment.

The vineyards of California consume annually 1,200 tons of sulphur, an average of about 15 lbs of sulphur per acre. None of this sulphur is the product of California mines or deposits as many suppose. For three years previous to 1887 sulphur sublimers and grinders have been entirely dependent on countries other than the United States for their raw material. There has of late years been no sulphur found in California which could pay the cost of mining, refining and transportation to San Francisco, and be sold here at even double the present cost of sulphur imported from Sicily or Japan.

Four years ago California received small quantities of sulphur from Nevada, but the competition of cheap labor in the Orient, and cheap transportation by sea soon choked out the local industry. Considerable promise comes to the home industry now from Utah, where large deposits are being worked, and the refined product, ground and sublimed, are being placed on the California market at the same figures as the imported vineyard sulphur ; or that prepared in San Francisco from the imported raw material. How long the sulphur mines of Utah will continue available to consumers in the United States will depend entirely upon railroad freights, which have of late been so capricious as to preclude any certain future dependence.

To show the comparative insignificance of our own sulphur mines, let it be known that in 1880 there were mined in the whole United States 600 tons, while our imports for 1881 aggregated 105,438 tons.

This latter quantity came almost exclusively from Sicily. Virtually, the Island of Sicily furnishes the world with sulphur, notwithstanding Japan is now her most formidable competitor for the Pacific Coast of North America. Sicily has for years been the original point of production for the sulphur used in vineyards the world over, and whether this sulphur sold mostly in commerce as "Sicily seconds" and containing not to exceed 3% of ash impurities, has gone first to Marseilles or Antwerp to be ground or sublimed ; or whether it has come to New York or California to be ground or sublimed, has made but one essential difference to California vineyardists, viz : All sulphur sublimed without the United States pays a duty on entering our posts of one cent per pound, which sometimes makes an addition of nearly 50% to the selling cost. All other brimstone, crude or ground, except in rolls, pays no duty.

I desire to draw particular attention to this difference for the instruction of those who have with this as with California wine in past years, been led to believe, that any goods bearing a French label are better than those produced at home. The case is a parallel to that by which our wine drinkers were long duped by French labels.

Some may claim that the sulphur ground or sublimed in Europe is finer than that prepared in California. To determine this I have examined carefully over twenty samples of sulphur which I have been collecting and carefully sampling for several years past, with the following results : Domestic preparations of sublimed sulphur have averaged as fine as those from Marseilles or Belgium. Of the ground sulphur, that produced in California has generally proved the finer, and the finest of all prepared by either method was ground sulphur prepared in California.

So much to the credit of the home industry. I have learned direct from the leading importer of foreign prepared sulphur that generally the Europeans do not grind as fine as is the practice in California ; but that if California markets so demand, it may be prepared as fine as needed. This is because they expect us to use sublimed sulphur if sulphur in a fine condition be needed. They care little about the import duty of one cent per pound on the latter so long as they are reimbursed. Nor is it their business or care whether we use one variety and avoid the tax, or the cheaper with equally beneficial results.

The imports of sublimed sulphur to a single merchant in this city have cost California vineyardists in the past three years nearly $15,000 duty, no benefits of which have accrued to our vineyards ; and this, a loss to proprietors, adds another conspicuous monument to the long and unwholesome practice among some of our people of aping the French in everything.

We therefore conclude from the foregoing that "California sulphur" means sulphur from other countries, ground or sublimed only in California ; and that for economy's sake, if any one insist on a foreign article, he should buy the ground sulphur and thereby escape the duty of one cent per pound.

COMPARATIVE VALUE OF GROUND AND SUBLIMED SULPHUR FOR VINEYARD USE.

This much mooted question has been carelessly handled by many. First let us comprehend the effect of any sulphur distributed in the vineyard. Sulphur, in a fine condition, exposed to the atmosphere, undergoes a partial evaporation ; the vapor produced comes in contact with the germs or organs of the oidium and accomplishes their destruction. Evaporation is therefore the result desired. This evaporation is particularly favored by exposure to the suns heat, and especially when the ambiant temperature reaches 70° F or over.

Now, other things being equal, the finer the sulphur the greater must be the surface exposed and consequently the more rapid the evaporation. Evaporation is the result desired. Sublimed sulphur is that produced by boiling crude brimstone and condensing the vapor thus formed in a closed chamber. In cooling the vapor the sulphur is recovered in little round globules. Several of these globules are usually attached, and form a string in appearance when magnified, much resembling a string of beads. The sulpur in this shape exposes less surface than could be produced in any other form ; so that with equal fineness, ground and irregular particles would better answer our purpose. To demonstrate this practically, two samples of the same sulphur accurately determined in weight, one ground and the other sublimed, were exposed to the same heat as if in the vineyard. Samples selected for this purpose were of apparent equal fineness. The same were weighed from day to day, and the experiment repeated several times ; and the above conclusions were amply born out in every weighing. Not only did the sublimed or flowers of sulphur evaporate less; but it also showed a more rapid formation of sulphuric acid than the ground sample, thus furnishing another objection to the use of sublimed sulphur ; one which its exponents have frequently and incorrectly urged against the use of ground or triturated sulphur.

I have yet to know of any considerable damage done to vines by the sulphuric acid existing as an impurity in the commercial article, either ground or sublimed ; though some have strongly urged the presence of sulphuric acid as an objection to ground sulphur. This is wrong, as there is every condition to favor the formation of sulphuric acid in the operation of subliming sulphur, and nothing to favor such formation in the grinding process. Specialists who have made this matter a thorough study, corroborated my conclusions.

I do not urge this as an objection to the use of sublimed sulphur, but if any disadvantage accrues from the presence of sulphuric acid, it must not be laid at the door of the ground sulphur.*

In connection with this let it be known that neither subliming nor grinding does in any wise alter the chemical nature of

*The presence of considerable quantities of sulphuric acid may be detected by the lumpy condition which results from its presence.

sulphur, which is an elementary substance and unalterable chemically, otherwise than by combining it with some other element or compound. It is not changed in its preparation as above named, any more than would be pure lead if made into shot by melting or by being cut to the proper shape. The same analogy holds true as to its source—pure lead from one mine or country is chemically identical with that from any other mine or country. So with sulphur from Sicily, from California or any other country. This I state for the benefit of non-chemists, some of whom have thought prepared sulphur to be a compound altered from its elementary condition and hence variable in strength.

One point favoring the sublimed sulphur is, that in preparing it, the product is freed of the ashy impurity existing in the crude article of commerce, to the extent of from one to three per cent. This impurity, however, is a neutral volcanic ash, which works no injury to the vine, and in buying ground sulphur can only be estimated as a loss of from one to three per cent.—a loss which is in no wise commensurate with the difference in price of the two forms, ground and sublimed.

We find European authorities of the present date unanimous in the opinion that finely ground or triturated sulphur is more suitable for vineyard use than the sublimed.

Prof. G. Foex, who is Director and Professor of viticulture at the National School of Agriculture at Montpellier, in his "*Cours Complet de Viticulture,*" published in 1886, says : "Formerly only sublimed sulphur was employed (in the vineyards) because it contained more sulphurous acid ; but since, learning that the effect of the sulphur on the oidium is due to the vapor which it emits at an elevated temperature, a result obtained as well with ground as with sublimed sulphur, the former being considerable cheaper, has come into general use.

"Furthermore, the use of flowers of sulphur is seriously objectionable, in as much as it consists of little globular par-

13

ticles which are readily lost from the foliage of the vine under the influence of light breeze. Its application affects the workmen, too, with a trouble known as *ophtalmie des sulfreurs* —affecting the eyes.

"The ground sulphur, on the contrary, which is made up of angular and irregular particles, adhere more closely to the green portions of the vine and trouble the workmen much less."

M. A. Du Boreuil, M. La Forgue and others express the same preference for finely ground sulphur.

HOW AND WHEN TO APPLY SULPHUR.

For very small vineyards, the dredger, an instrument much resembling a large pepper box, answers well enough, especially while the vines are young. For more advanced vineyards and larger areas, the bellows should be used, holding from three to five pounds of sulphur. These latter are furnished by local manufacturers and effect a considerable saving of time, labor and material over the dredger. By the use of the bellows, too, the sulphur can be more evenly distributed. A simple open nozzle is the best ; any perforated cover for this latter is apt to get clogged and the bell-shape frequently given to it does not spread or expand the sulphur jet—a purpose for which it is designed but fails to accomplish. A bent nozzle is more of an encumbrance than an advantage. The simplest, strongest bellows of good size will prove cheapest and best in the end. With this tool a workman will sulphur from five to eight thousand vines per day—vines in an advanced state of vegetation. He may apply as many pounds of sulphur per day with other instruments, but it can not thus be so evenly distributed, nor cover the same area.

The powdered sulphur should be applied so as to lodge as much as possible on and near the growing parts of the vine. This secures a dense sulphur vapor in direct contact with the diseased organs. Sulphur on the old stump, or even on the surface of the ground, will destroy the oidium, but a larger quantity would be required.

Sulphur falling on the ground is by no means lost, but a lesser quantity will answer if lodged on the leaves and branches. It has been stated that sulphur falling to the earth is lost by its effect being immediately neutralized. The sulphurous acid formed is neutralized, but the vapor of sulphur—the active disinfectant is not neutralized, nor is the effect of the sulphur lost, except as it be covered up and hid from the sun and air.

The simplest rule as to the time for applying sulphur is: "Treat the vineyard whenever the disease makes its appearance." But if we desire to apprehend even its introduction, which is the general custom in California, the first application should be made at or about the time of flowering, as at this period the disease is apt to attack the delicate organs of fructification and render the vine infertile. Altogether the most favorable results have been obtained by sulphuring at the time of blossoming. This, too, is one of the methods of combating *coulure*, a trouble which will be treated later on. Young vines do not require so frequent sulphuring nor so great a quantity as vines in full bearing. The former should be sulphured when the shoots attain a length of a few inches; and again, later on, if the oidium makes its appearance. Bearing vines should, in addition to the treatment at blossoming, receive a second application from the first to the middle of June, and again, later on, if the disease makes its appearance.

The quantity used at each application may vary with the number of vines per acre, and should be governed somewhat by the susceptibility of the variety, as before explained. Less is needed for the first than for subsequent applications, when the vines attain full proportions. The quantity commonly used in California for old vines subject to oidium, is about eight pounds per acre for the first treatment, and from twelve to twenty pounds at the second application.

The use of this remedy in conformance with the above instructions will affect a great-saving over conventional methods pertaining at present in California. Not unfrequently our vineyardists sulphur in weather positively prohibitory to the disease, Varieties but little liable to oidium, situated perhaps in the hottest and driest interior localities and trained low to escape it, often receive a dose which goes only to fertilize the soil and stimulate the growth of the vine.

This latter function is one which, however, must not be overlooked; the general aspect of the vine is always improved and vegetation greatly stimulated by the free use of sulphur. A small percentage only of the sulphur applied vaporizes—the balance works into the soil, becomes slowly oxydized and finally unites to form sulphates of the alkalies and alkaline earths, which are in substance the essential ingredient of some of the best fertilizers. Still it is well to know whether the sulphur is applied for the cure of oidium and as a fertilizer, or as a fertilizer only.

The most favorable hours for applying sulphur are from eight or nine o'clock in the morning to the middle of the afternoon, preferably from 9 A. M. till 2 P. M. The sulphur which comes in contact with dew or other water is in no wise altered thereby, but ceases to give off its vapor only until the water evaporates, and thereby exposes its surface to the atmosphere. A rain following the application of sulphur does not alter this element, but results in damage only in removing the particles mechanically from the foliage of the vine to other places more remote from the seat of disease.

Any wind other than very gentle will do much to shake off and remove the sulphur from the leaves. A windy day should therefore be avoided. In fact, a hot, still midday is best in all respects, as amply proven by the strong odor of sulphur prevailing at the time of such an application.

In purchasing sulphur, its quality and fineness may best be determined by the use of the microscope. The weight of a given bulk will establish the relative fineness, but cannot be used in comparing sublimed with ground, as the mechanical condition of the two are different—sublimed sulphur is in beaded strings and occupies more space than ground sulphur, much as shavings are of greater bulk than sawdust. To persons experienced in its use, the fineness may be determined by the feeling — almost impalpable it should be. I have never known this sulphur adulterated, although ground sulphur is frequently added to the sublimed to enable the merchant to sell it cheaper. All samples of Eastern sublimed sulphur examined I have found mixed in this manner, showing that the credit of ground sulphur has sometimes been unconsciously extended to the so-called and more expensive sublimed.

Where the question of purity or fineness arises with any vineyardist, samples may be sent to this office, where a prompt determination will be made and reported without cost to the applicant.

To further substantiate these recommendations of the ground sulphur, I will state that of the sulphur used of late years in California vineyards, over three-fourths has been ground sulphur; and I have yet to know of anyone employing sulphur extensively who has abandoned the ground, or even that ground in California, for the foreign prepared or sublimed sulphur, which latter sells at one and a quarter to one and a half cents higher than the ground or triturated.

Ground sulphur may be easily distinguished from sublimed by the difference in color, the latter always shows more yellow, the former more white or a lighter tint of yellow.

Some varieties are constitutionally and inherently inclined to this disease, for instance the Malbeck, which always suffers from it, and therefore is not worth cultivating here. In others, it arises from unfavorable conditions of the atmosphere, improper location and soil, etc. Among those most easily affected by these changes, I note the Zinfandel, Muscat of Alexandria, Flame Tokay, Cabernet Sauvignon, Chaucne Noir, the Rieslings, Gelber Mosler, White Elben, etc. With these, it generally follows sudden changes of temperature; for instance, very cool weather or frost, followed by hot northern winds; windy locations are more subject to it than those somewhat protected, etc. Sometimes, late rains, which keep the ground cold and moist, are the cause ; and again, methods of pruning. If the vine has been pruned too short, rank growth and coulure are almost sure to follow; and vice versa, if the vine is taxed beyond its strength, it is apt to set imperfectly. While we may not be able to control atmospheric influences entirely, we can certainly do much by pruning properly, taxing the vine neither too much nor too little; and it also should lead us to be careful in choosing our location, avoiding cold, damp soils and exposed locations. No doubt we can also prevent it to a certain extent by using certain fertilizers, applied just before the bloom. Foremost of these, where it can be had, is the ammoniacal liquid, which can generally be had at gas works for very little more than filling into casks, and which is the cheapest and best fertilizer I know. Dilute each gallon of the liquid with seven gallons of water, put a cask on a low wagon or sled, and attach a hose with a faucet to it. Make a small hole above each vine, say eighteen inches or two feet from and above it, and apply about half a gallon to each vine. Two men can go over about five acres in a day, and it is wonderful how it stimulates the vine and increases the size of the berries, while the cost is mainly in the application. Sulphuring at time of bloom is also a partial preventative, whether its evaporation counteracts the deleterious atmospheric influences, or through its action as a fertilizer, or both, I do not pretend to decide, but there is no doubt in my mind as to its beneficial results. Early pinching or summer pruning before the bloom is one of the main preventatives. I think I am also warranted in saying, that grafted vines are less subject to it than those not grafted; and as French authorities recommend girdling of the vine or shoot as a preventative, I hold that grafting, forming a temporary obstruction to the flow of the sap downwards, has a tendency to make the vine set better. We know that grafted trees of any kind set their fruit better than seedlings; it is reasonable therefore to infer the same of the vine. Ringing, or twisting a wire temporarily around the cane or shoot early in spring, will accomplish the same result, but this is a laborious process, and also apt to injure the vine. Binding and twisting the canes in long pruned varieties, or bending them in a circular hoop, will also tend to prevent coulure.

Some very interesting observations on coulure in San Diego County, by Mr. F. G. Morse, of the State University, can be found in the Report of the Viticultural work of 1885–1886, by Professor Hilgard, to which I refer those of my readers who wish to inform themselves further.

Red Leaf, (Spanish Measles), Anthracnose, Pocken des Weinstocks. Whether what we know by the two first names is identical with what is known in France as *Anthracnose*, in Germany as *Pocken des Weinstocks*, I am not quite sure, but presume they are identical. It generally appears about midsummer, and I have mostly seen it on old Mission vines, which had been pruned to the stool shape for quite a number of years, and on which the saw had been formerly used. The disease often attacks but a single spur, sometimes half of

the vine, sometimes it takes all, while in another season it
may be healthy again, ripen its fruit, and have healthy leaves.
The leaves become spotted, like if drops of hot water had
fallen on them, becoming livid at first, they soon change to
almost scarlet, and finally drop; the fruit, if small, shrivels
up and dries; if larger, it becomes marbled with dull gray,
and does not attain full size. I think some injury to the
vine, either by injudicious and severe pruning, or by tearing
the roots with the plow, or by gophers or other animals,
mostly the cause of it. The bluestone remedy has been
recommended against it, and may prove effectual. It is prepared as follows: Dissolve in 25 gallons of water 16 lbs. of
copper sulphate (bluestone), also shake 20 lbs of quicklime
with seven and a half gallons of water into a milk of lime,
which then rinse with the bluestone solution; this will produce a light blue mixture, which should be frequently stirred
during use. This is sprinkled on the vines by means of a
little broom. Care should be taken not to sprinkle the
grapes. It is recommeded by Professor Millardet also as a
remedy against peronospora, and Prof. Hilgard thinks it may
be a remedy against mildew. I think that an application of
this solution, applied as a whitewash to the body of the vine in
the winter, would be very beneficial as a preventative of
diseases, and also destroy a great many injurious insects and
their eggs.

BLACK KNOT.

This disease, which appears sometimes on the trunk of the
vine; sometimes on its head or below a spur, is mostly due to
external injuries, either too short or injudicious pruning,
bruises, breaking of the vine, or severe frosts; in short, by
some cause which effects a bursting of the sap vessels. For
instance, if all the young growth of the vine is so badly frosted
that even the wood is affected, and the dormant buds killed,
there is nothing left to conduct the flow of sap, and stagnation

ensues, by which the sap vessels below are extended to bursting, the sap vitiates, and in oozing out through the bark, forms these abnormal warty excrescences. Vines grown from cuttings of very large, porous wood are also more subject to it than those from medium, firm, short-jointed wood. I have already referred to injudicious short pruning, reducing a vine of say 300 buds, to eight or ten, as one, and the most prevalent, cause of black knot. Judicious pruning, and in case of very destructive frosts in Fall or Spring, or the breaking of the vine as may sometimes happen, grafting may prove a preventative, as the scions will then serve as conductors and elevators of the superfluous sap. In resorting to this remedy in large vines, it will be well to take them low down, and leave several buds to the scions.

CHAPTER XVI.

INSECTS AND ANIMALS INJURIOUS TO THE VINE.

Our most formidable insect enemy, the phylloxera, has already been discussed in a previous chapter. Other insect enemies are not so formidable, though sometimes injurious enough. Perhaps the worst of these is the little white thrip, a leaf hopper, a little midge of a thing, which feeds on the under side of the leaves, causing them to dry or drop, when the fruit can not fully ripen, and therefore will not develop sufficient sugar. Early summer pruning is one of the aids to prevent exposure to the sun, as the young laterals will retain their leaves much longer. But sulphuring with bellows will also serve as a partial remedy, and kindling small fires in the vineyard at night, or one person taking a brightly burning torch and walking through the rows, while another beats the vines, will cause the insects to fly into the flames and thus get scorched.

The Grape Vine Fidia, a small ashy gray beetle, about the size of a common house fly, sometimes becomes very destructive to the foliage. Sulphuring, and when they become too numerous, hand-shaking early in the morning, when the insect is still dormant, into a screen of the shape of an inverted umbrella, with a slit or space on one side to enable the operator to push it under the vine, are about the most common remedies.

The gray cut worm and the wire worm, a worm about two inches long with a hard covering of brownish yellow color, sometimes materially injure the young shoots. The wire worm works mostly underground, while the cut worm will cut off the young shoots above the ground. Handpicking is about the only remedy; the cut worm is generally found under the loose clods at the base of the vine, while the wire worm is found mostly on the suckers below and the young shoots of the grafts where they are below the ground.

There is also a black, longish beetle which will bore into the buds and wood, making a round hole, but I have not found it very numerous or very destructive. A steel blue beetle, very active, is also destructive to the young shoots, and sometimes a large worm, similar to the common tobacco worm, will feed upon the foliage. The leaf folders common in the East I have not yet observed here.

The Rocky Mountain locust, or grasshopper, has visited certain sections of the State, and is very destructive. I have seen a vineyard of one hundred acres in Knights Valley kept completely bare by them one summer ; but in the next they had entirely disappeared. It is very difficult to guard against them, though the remedies advised by Prof. Riley can no doubt be of some use, digging ditches and then crushing them, etc. There visitations, however, are very few, and seem to be only temporary, as the rains of the next winter, together with their insect enemies, have destroyed them.

It seems also that they destroy all other insect enemies to the vine, or starve them out, as seldom any of them are seen the next summer.

Bees and wasps are sometimes quite troublesome, and it is certainly not advisable to grow grapes and also keep bees. Traps of small jars, filled with a solution of molasses, into which they will crawl and drown themselves, is about the best remedy.

The punctured Diabrotica, a small beetle of the size of a common lady bug, also preys upon the berries, eating holes into them. So far it has not been very destructive. Our common lady bug, the little red and black beetle, is accused by some of feeding on the bloom and the young berries, but I think erroneously. I have always considered it one of the best friends of the vineyardist and orchardist, as they destroy thousands of aphis or plant lice, ants and thrips, and I should be very sorry to see my little friend convicted of real mischief.

Among our best friends we may also count the common toads and lizards, which destroy countless insects, and should be carefully preserved and fostered, not tortured and killed, as thoughtless children will do sometimes.

I believe that a solution of London purple, about one pound to fifty gallons of water, sprinkled over the vines before the bloom, will destroy most of the insects that prey on the foliage. I trust that we will soon have conclusive evidence as to its merits.

Rabbits or hares are sometimes very destructive, biting off the young vines and grafts. Other remedies, such as blood, etc., smeared on the vines, have done but little good so far. When they are numerous, a tight picket fence is the best safe-

guard, such as is now woven by machines at 80 cents per
rod, with five double wires, which can be fastened to posts by
staples, and is an effectual protection.   Otherwise the shot
gun and grey hounds are the best protectors.

Ground squirrels and skunks, also raccoons and foxes, are
all very fond of grapes.   The best remedy against the first,
and which also generally tells on the others, as they will eat
the poisoned squirrel, is Mc'Leods squirrel poison, made at
Livermore, Alameda Co.   It is poisoned wheat, flavored with
Angelica, the smell of which seems to draw the squirrels, and
is instantaneous death to them.   We have killed hundreds of
them by a single application of four or five grains, thrown into
their holes.   They can generally be found the next morning
in front of their holes, and should be looked after and buried.

The pretty little California quail, although no doubt very
useful during the summer in destroying insects, becomes a
great nuisance in fall, and I think it was wise in the super-
visors of Napa County to change the season of their protection from 1st of October to 15th of August, as vineyardists
can now use the gun against them in time to reduce their
depredations.   When we consider that they live entirely of
grapes as soon as they ripen, and that they will use the "grape
cure" at the rate of a bunch a day for each, we can easily
imagine what an expensive luxury they may become.

CHAPTER XVII.

FROSTS AND HAIL, THEIR EFFECTS, PREVENTATIVES AND REMEDIES.

While many sections of the State are free from frosts, others, and among them those which produce our choicest wines, are very much subject to them, and they have proved so capricious of late that it cannot be said that any location in those sections is entirely safe. Locations which had not suffered from them for ten years were frosted last spring, while others which were frosted badly in former years, escaped unhurt this season. I speak here of late spring frosts and early frosts in fall. Our winters are not severe enough to hurt the vines, unless in abnormal seasons, where a moist fall which started the vines into an unnatural growth and they did not mature their wood fully, was followed by a sudden snap of cold weather. This is so seldom the case, however, that it should hardly be taken into consideration, and only in low moist locations not fit for grape culture; which should be avoided in planting anyway.

But while we cannot say that any locality in some sections of the State is entirely free from frost, yet there is a great difference. Low, narrow valleys and springy ground are peculiarly subject to it, and should therefore be avoided in choosing a location. Very often a few feet of elevation will make a great difference, and the vines in the valley may be all black with frost, while five feet above it, rising towards the hills, not a leaf may be touched. Therefore avoid low, moist locations; these will do, if you have them on your place, for grain and hay, vegetables, etc.; and plant your vines on the

warm hillsides, sloping down to the valley, which will give
you a choicer product than the rich valleys anyway, though
perhaps not quite so much in seasons free from frost. It is
discouraging to see a vineyard, rich in promise of a bountiful
crop, cut down and blackened in a single night, although the
damage is seldom so great as it appears at first sight.

Then plant your lowest blocks with such varieties as will
start late and bear well even from the lower buds. The Marsanne, Green Hungarian, Pedro Ximenes (generally called
Sauvignon Vert, Colombar erroneously) Palomino (Golden
Chasselas, erroneously of Napa), Clairette Blanche and Mataro
are safest, and will yield a good crop even if the first growth
is frosted, under a treatment I shall describe later. So much
in regard to locating and planting as *preventative* measures.
Among the other mechanical preventatives we will consider

1st. Late Pruning. This is advocated and practiced by
some, who argue that by deferring the pruning until May, when
the vine has already grown six inches or more, they can keep
the lower buds on the shoots dormant, as the upper buds start
first, and if they wait until danger from frost is past and then
prune, they will have a crop from the lower buds. This is
no doubt true, but the vine must also necessarily receive a
severe shock, and be enfeebled thereby, especially as they
bleed very severely if pruned at that time. It also makes the
fruit very much later, there is danger of imperfect ripening
and immature growth in fall. Moreover. it is a great waste
of energy, to allow the vine to produce so many shoots and
then prune them off. For all these reasons I do not think
the practice should prevail.

2nd. High Training. Some grow their stool pruning varieties with heads four and even five feet high, claiming that
they suffer less thus then when pruned to low heads. This may
be sometimes the case, but in seasons like the last we have
seen that in the same piece of vineyard, sometimes the upper

shoots were killed, while the lower escaped; while a neighboring vine would show exactly the reverse; many had their lower and upper shoots killed, while the middle ones escaped; and another vine close by was not hurt at all. Therefore in such seasons high training would not be a preventative, although in some cases it may prove beneficial; but we can hardly expect as good results from vines thus elevated, nor quite as early.

3d. Smoking. This is one of the most generally adopted preventatives. It is claimed by its advocates that by raising a dense smoke early in the morning, about three to four o'clock, it will raise a cloud or covering above the vineyard which will prevent the effects of frost, if any comes, and moreover, prevent the direct rays of the sun from striking the vines, should they be frosted ; and thus allow them to thaw gradually. This is a plausible theory, and may hold good to a certain extent. If all the neighbors join and make so dense a smoke that it will serve as a heavy cloud over all the vineyards, and the temperature does not fall too low, it may prevent or ameliorate damage. But they seldom work in unison, some prefer staying in bed while others watch and smoke; besides, I think smoke is only a preventative when the thermometer falls to the freezing point or slightly below it; but when it drops as low as 28°, or even 26°, as has been the case, even the most systematical smoking will not save the vines from damage. But if all these preventatives will fail sometimes, what shall we do to be safe, or, at least, partially so ?

I have given this subject close attention and thought, and I think my experience of last spring has proven that I have found a method to obtain fair crops, even in the most frosty season. My method and partial preventative is "longer pruning."

Instead of the commonly followed practice of pruning all spurs to two buds, I leave four, otherwise pruning as before ;

leaving just as many spurs or canes as usual, but also leaving the canes somewhat longer than usual. Now for the result.

As part of my vineyard is in the valley and on very rich soil, where the vines make a very vigorous growth, I pruned them last winter as indicated above, fearing there might be trouble from frost. Last spring (1887) was a very frosty one, and my vineyard was visited by Jack Frost two distinct times, once the latter part of April, the last and most destructive one being May 12 and 13. On some of my vines, which had started vigorously, with shoots already 18 inches long from the upper buds, not a green leaf was left to tell the tale; while others had the tops of the shoots badly blackened, and some escaped with little or no injury. The vines presented a truly sickening and discouraging aspect, and my two sons gave up to one-fourth of a crop. As soon as we could ascertain the full extent of the damage, we armed ourselves each with a small pair of shears we use for picking grapes, and cut all the blackened shoots off clean; while those which had only the tips slightly singed, were cut back only as far as damaged. It cost three of us a full week's work to go over 20 acres of bearing vines, and when the job was done, about 5 acres presented a perfectly barren appearance, with only here and there a green shoot or a few leaves left. My readers will please bear in mind that all the frozen shoots had started from the *upper buds* on the spurs as well as on the canes in long pruned varieties. The varieties were Zinfandel, Chauchè Gris, Chauchè Noir, Franken Riesling, Pedro Ximenes, (Sauvignon Vert as known in this State), some Madeline Angevine, Orleans Riesling, Mataro. Of these the Pedro Ximenes and Mataro were only partly damaged, while the others fared about alike. The vines remained in this almost dormant condition for about a week, when the lower buds, dormant so far on the spurs as well as the canes, commenced to grow, and in about a week more they were all out in leaf, showing abundance of fruit forms, except those shoots which we thought had been injured but slightly. These attempted to bloom, but as we had a few days of hot north wind just at that time, they dropped nearly all their bloom. At the present writing (August 15), the vines that were entirely frosted show a nice crop of about 3 tons to the acre, of finely set and developed bunches and berries, while those only partially frosted are not near so good, with the exception of the Pedro Ximenes and Mataro, which show a good crop from first growth. The vines have made an enormous growth, were pinched when the young shoots were 18 inches long, and are now interlaced with each other, so that it is difficult to get through between.

The conclusions to be drawn from this are very simple. When there is any danger from frost, we should prune to double the amount of buds, to be safe. If no frost comes, we can easily rub out all weak and superfluous shoots, reducing them to one half the number. This will give them all more room and air, and it is certainly easier to rub off a superfluous shoot, than to add to those which nature has alotted. Pruning is rather a perversion of nature any way; and when we overstep the bounds of reason, when we infringe her laws by mutilating, instead of only reducing to such limits as will give us the most perfect fruit, outraged nature will rebel and punish us in time.

Mr. John H. Wheeler, our present chief viticultural officer, has experimented in the same direction, and as he tells me, with favorable results similar to mine. The difference in his treatment from mine only consists in his leaving still longer canes, and not so many spurs.

After this season's experience, perhaps one of the worst we have had, in its sudden changes from low temperature to scorching north winds, I have little fears of raising a satisfactory crop *every* season, even after destructive frosts. Still, I

14

shall extend my vineyards to the hillsides and more elevated slopes, of which I have an abundance, and would advise every one to do the same. I would not have planted in the valley, but my predecessor thought himself safe, as there had not been any frosts there within the memory of that famous personage, the "oldest inhabitant."

Sometimes, but very rarely, we are visited by destructive hailstorms in early spring, which are similar in their effects to frost. I would advise the same treatment in that case, cutting back the injured shoots, and trusting to the dormant buds for the crop.

Yet another point in this connection ought to be mentioned here. It is often the case that we have extremely hot weather at the beginning of the vintage, when fermentation sets in with great violence, but stops at a certain point, which I shall more fully explain when I come to wine making. Between the 15th of September and the 1st of October, we generally have a few showers which refresh the vines and the grapes; after which there seldom is any trouble, but fermentation goes on normally and well. This later crop, delayed about three weeks, will be apt to escape this trying period, ripening after it, but still with abundance of sugar to make a first class wine, and I think will have more bouquet and sprightliness than the first would have had. It is not what is termed "second crop" which comes on the laterals of the fruit bearing shoots, but "first crop" from the main shoots, only delayed a few weeks. I hope to report to my readers, before this volumes reaches them, how far these conjectures have been verified by the facts.

Some varieties are also much hardier than others, and among these I will name the Pedro Ximenes (erroneously Sauvignon Vert, Colombar) Mataro, Marsanne, Green Hungarian, Clairette Blanche. All of these start late in spring, and are therefore not apt to suffer so much.

Since the above was written we have approached the vintage, a full month having elapsed, and I can report now, September 16th, a larger and better crop than even predicted then. The totally frosted vines will make four to five tons to the acre, of fine compact bunches, which will all be fully ripe by 1st of October, and which certainly will make a first-class wine. So fully am I convinced of having found the best remedy and preventative against frost, that I shall not attempt to smoke, but trust entirely to long pruning; and feel certain of a crop.

CHAPTER XVIII.

THE VINTAGE—GATHERING THE GRAPES FOR WINE..

For this, the grapes should be thoroughly *ripe*, yet not *too ripe*. There is a period in the maturity of every fruit, when it is at perfection; as soon as this period is passed, it approaches decay, loses sprightliness, and while it may develop more sugar, and, as is the case in the grape, turn into raisins, its wine will lose in freshness and bouquet, and gain only a larger percentage of alcohol. If, on the contrary, the grapes are picked too green, the wine will always have a greenish, unripe taste, and be harsh and sour, owing to the surplus of tartaric acid and malic acid.

General indications of ripeness in the fruit. The stem of the bunch changes from green to brown, between the shoot and the small knob on the stem above the bunch, and the bunch becomes pendant. The berry becomes translucent and soft, its skin thin, and they separate easily from the stem.

The juice acquires an agreeable sweetness and flavor, and becomes thick and glutinous. The seeds separate easily from
the flesh.

These will serve as general indications, but the surest test
is the sacharometer or must scale, of which there are many
kinds; but as Balling's is the one commonly used in this
State, and is about the simplest and best, we will take it as
our guide. They are all constructed on the same principle,
that of the density of a fluid, for instance water, being increased
by adding sugar; and therefore the sweeter the must, the less
will the instrument sink in it, or rather, the more will the
sugar uphold it. If an average sample of grapes is taken
from the vineyard, the juice expressed and strained, and
at the right temperature, which should be from 62 to 65°
Farenheit, the instrument will test it with certainty, and is
surer than all other indications. Figures 23 and 24 will show

Balling's Sacharometer and the testing
jar, which can be of glass or tin, with
a wider foot or rim to stand upright.
If none of these are at hand, any jar
or glass that has the necessary depth
for the sacharometer to sink, will answer. Fill the jar full enough so that
when the instrument sinks in it, the
fluid will be close to the rim, then
wipe your instrument carefully, holding it by the stem, and let it sink
gradually until it floats. Then press
lightly with the finger, so that it will

FIG. 23.   FIG. 24.

come to an equilibrium, but be careful that there is no fluid
on the top above, as that would influence its accuracy.
The surface of the liquid, when the instrument has become
stationary, will indicate the sugar contained in the must,

where it touches the number on the stem of the instrument. The must should not be less than twenty-two or higher than twenty-five, to make a good, light white table wine or claret. But there is a certain difference in varieties, which should be borne in mind. Some varieties ripen irregularly, and have quite a number of overripe berries; while others are hardly ripe enough. The Zinfandel is one of these, and as these shriveled berries contain hardly any juice, but rather dried sugar or caramel, the sacharometer shows less sugar in these than they really contain. The Zinfandel should be *very ripe*, to make as full flavored and smooth a wine as it will, if properly handled. In testing such grapes, always count from 1^{U} to 2^{U} *more*, than the instrument shows. If such Zinfandel must show 23^{U} count it 25, and your grapes are ripe enough. If it does not show this, wait with the vintage until it does, except in abnormal seasons, when they will hardly come up to it, and yet may commence to rot, which is a sure sign that they should be picked. In Burger again, and a few others, you can hardly expect more than 20 to 22°, even when very ripe, and it would be unwise to wait longer. The remedy in these cases will be to blend with a heavier wine, although 20^{-} being equal to about 10 per cent. alcohol in the fermented wine, is really heavy enough for an ordinary light wine.

In this connection, let me impress upon my readers the importance of a succession of varieties, from early to late, so that they will not be crowded by one variety which ripens at the same time, and of which they would be forced to pick some hardly ripe enough at first, while the last would be over ripe. Have a succession of, say six varieties in about equal quantity, so that you can take a week for each, and do it justice. There is one great trouble, with which large wineries, who purchase grapes in addition to their own product, have to contend with; they cannot control the ripeness of the grapes, nor the supply each day, as well as the individual,

who works up only his own product; they are often rushed
by their supplies beyond their capacity, and the consequence
is a great deal of hasty and imperfect work, resulting in
faulty wines.

IMPLEMENTS FOR PICKING.

Knives are generally used, but I find a small scissor or
shear, imported by Justinian Caire, and manufactured at
Geneva, much handier.   Fig. 25 shows the implement about
two-thirds natural size.   They are very
convenient, cut easily, can be held in the
hand without strain, a plated spring opens
them wide enough for cutting, and they
do not shake the bunch and vine as a
knife will, thereby preventing dropping of
berries and wastage.   Their price is about
$7 per dozen, or 75 cts. retail.   Large
pruning shears can also be used, but are
not near so handy.

For receptacles of the grapes we generally use wooden boxes here, holding
from 40 to 60 pounds each, with oblong
holes in the ends, so that the workman
can carry them before him.   They are
mostly bought in shooks, nailed together
at home, and bound at the ends with a
strip of rawhide or wire, to keep them
more solid.   Care should be taken that
the bottoms are of one piece, as it prevents breakage and waste through the
cracks.   They are, when filled, carried
to the avenues, from where the wagon
takes them to the winery direct.   When
grapes, especially the tender skinned
varieties, must be hauled a long distance

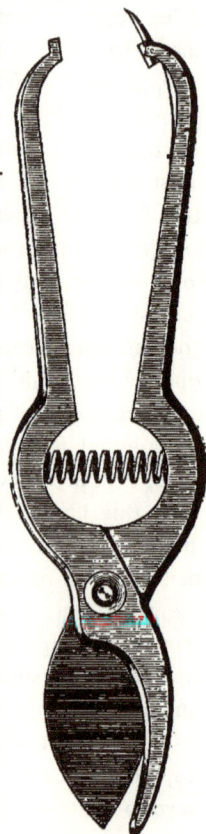

FIG. 25.

over rough roads, it will be found advantageous to have two oblong vats, four feet high, made of good redwood, and to fit on an ordinary wagon, into which the boxes can be emptied. This will prevent leakage, and the juice which runs out and is wasted, is generally the best ; the grapes can be pitched with forks on to the elevator at the winery, and thus save it all.

The number of pickers to be engaged, depends altogether on the quantity to be worked up each day. A man can pick and carry from 25 to 50 boxes, say at an average, a ton a day, which will vary with the varieties of grapes, whether large or small bunches, etc. What is picked in one day, should fill at least one vat or cask when pressed, and be crushed the same evening or day. The grapes should also not be too warm; if the temperature rises to 95 or 100 as it sometimes does for a few days, they should be allowed to cool during the night, if necessary ; and worked up in the morning. Of this I shall treat more fully in " Making the Wine."

It will become necessary sometimes, when the grapes do not ripen evenly, from the influence of frosts or other causes, to pick several times, taking only the bunches that are fully ripe. Pickers should be closely instructed and watched in this respect, so that only evenly ripened grapes, which alone can make a good wine, are taken. They should also be instructed to pick out all decayed or rotten berries, as well as those affected by mildew and Red Leaf. All of these are apt to introduce the germs of disease into the wine. Sound, finely flavored wine can only be made from perfectly sound grapes, well ripened; and any negligence of this kind will retaliate on the wine maker.

It will sometimes be found desirable to blend several varieties, and when they ripen at the same time it is best to blend by gathering at same time and mixing in the fermenting vat. For instance, the Chasselas Fontainebleau (Gutedel) and Victoria Chasselas will blend well together, as the Victoria takes away the softness of the Gutedel, and gives it more character. For this purpose, and to find the best blends, the winemaker should experiment, and he will soon find how to blend, and in what proportions. No rules can be given that will apply everywhere and in every season, as the product will change with location and season.

It is also necessary to consider what class of wines are to be made from the grapes. If light, dry wines, with fine bouquet and sprightliness are desired, the grapes as before remarked, should be ripe, but not over ripe. If very full bodied, smooth wines are the object, let them get fully ripe. If for sweet wines, let them remain as long as they can hang, without decaying.

Boxes and all other utensils should of course be perfectly clean and sweet, so that there is no danger of acquiring a mouldy or impure taste.

CHAPTER XIX.

PICKING THE GRAPES FOR TABLE AND MARKET. THE GRAPE CURE.

The information I can give on this important branch of grape growing is very limited, as my attention has been devoted almost entirely to wine making. I must, therefor, refer my readers mostly to other sources for details of the business, packing, etc. The most common package now in use is the square basket holding about six pounds. The grapes are picked the day before, so that the stems wilt slightly, which makes them pack more solid and convenient. The bunches are then carefully assorted, all imperfect or decayed berries clipped out with a small scissors, the bunches divided if necessary, to pack more conveniently and snugly, and then shipped across the continent. This season's shipments to New York and other Eastern markets seem to have been very successful in the main, and realized good prices. The first and earliest come from Vacaville and Pleasant Valley, while the great bulk seems to come from Sacramento Valley, and the latest from the Santa Cruz Mountains, where they remained fresh and green on the vines until January last year. The following is an extract from a letter of Mr. William B. West, of Stockton, and contains some interesting data:

"As to the raisin interest, I am not up to the times. I have found long ago that the Muscatel cannot be raised here, and I gave it up. In the shipment of table grapes we find that many good varieties, which were formerly considered too tender to bear shipment to the East, with improved cars and manner of packing, arrive in good condition. The soil and climate of the vicinity of Stockton has proved to be unusually good to produce a hardy grape that will ship well, and in the future this will probably be our chief production. Our grapes are not early; but when they are ripe they are a formidable competitor in the market. The first good grape that we have, which ripens about the middle of August and continues a month, is the Black Prince, or Rose of Peru; it is a very, firm, sweet grape; many tons are shipped to Utah, Montana, and Texas, from Stockton. Next is the flame colored Tokay, which colors well here, and is a very superior grape for long journeys as the season advances. We also ship the Mission, which is very different here from those grown at Napa, being much larger and sweeter. We shipped several loads of them last season, with good profits, but they were very fine.

"Still later, we have the Black Ferrara, a fine large grape with a deep blue bloom, a very abundant bearer, usually ten to twelve tons per acre, and an excellent keeper. It ripens from the 5th of October to November. The Emperor is also a great favorite in the East; it is a deep ruby red, keeps well, but is not so regular and good a bearer as the Ferrara.

"These are our most reliable shipping varieties; of course we use the early varieties, such as the Sweetwater and Chasselas Musqui, but as even our earliest kinds are more backward than those of Vacaville, they are not so profitable as later and finer kinds.

"I have fruited the Loja, or grape of Almeria, the variety sold in the East from Spain. I find, however, that it requires a more moist and sandy soil than mine; it does better upon a river-bank vineyard about ten miles from Stockton. Mine have good keeping qualities, but are not very large or prolific. I think this variety should be grown in a warm locality, where the roots could have an unlimited supply of water, as it has at Almeria, Spain. There are several Spanish varieties which do well, but they are white, and not much sought after here or in the interior.

"I did not mention the Muscat of Alexandria, as it is not grown successfully in our strong soil. Some day the Spanish kinds will be grown for shipment to the East; but we must find a very warm locality. Stockton, notwithstanding the generally prevailing impression, is not warm; our springs are cool."

The Natoma vineyards at Folsom have been very successful in shipping grapes from their locality, and always realized the highest prices. Valuable information will be found in the chapter "Individual enterprise" under the head of Natoma vineyards. Sonoma county has always been successful in producing fine Tokay grapes, and as the prices for good shipping grapes vary from $40 to $65 per ton, delivered to the packer, this has been more profitable than wine grapes. But in this, California has entered an entirely new and untrodden field, has had to learn solely by its own experience, and is only now beginning to see its way to a very successful trade. It will be some time before we are fully posted as to best varieties and localities adapted to them, methods of lengthening out the season, etc. One very important step is the establishment of four cold storage rooms, of a capacity each of 30x35 feet, in one of the wings of the old sugar refinery at San Francisco, now occupied by the Wine Storage and Security Co., San Francisco. It is the intention of the Company to keep grapes and other perishable fruits here all winter, and I hope they will meet with all the success their enterprise de-

serves. They seem to spare no pains to meet all the requirements of complete cold storage. No doubt fresh grapes, well
kept, would bring a good price in February and March.

For home market, they are generally packed in 20℔ boxes ;
and all methods of packing require, as a matter of course
great nicety and care in handling the fruit. So far, it has
been done mostly by Chinese help, but I do not see why female help could not be used as well, as has already been done
so largely in canneries, and with such entire success. Surely
this is work to which their nimble fingers, and taste for the
beautiful should eminently fit them. A great many grapes
are also canned every season, in the common quart cans. For
this purpose, the Muscat of Alexandria is used in preference,
on account of its color and fine muscat flavor, the largest and
most perfect berries only being used for the purpose ; while
the smaller berries can be utilized for grape jelly.

But one very important method of utilizing the grapes seems
to have been strangely overlooked and neglected so far ; the
grape cure, so largely followed and universally recommended
by the leading physicians of Europe. Yet it would seem to
be more needed on this dyspeptic continent than anywhere
else. Thousands upon thousands flock annually to the
vineyards of the Rhine, the Moselle and the Danube ; they
commence with eating half a pound of grapes per day, which,
before a month is over, is generally increased to four and five
pounds daily, and is considered the universal remedy for
impaired digestion and diseases of the bowels and kidneys. Is
it not strange, that here, where so many suffer from these diseases, and the remedy is at their door, so to say, it is used so
little, and seldom recommended by our physicians? Let me
suggest to the landlords of our numerous summer resorts the
propriety, nay the necessity, of having at least a few acres of
vineyard connected with their establishments, where there
guests can have fresh grapes at any time during the autumn

months ; fresh and cool every morning, with the dews of the
night still sparkling on them.   I venture to predict that their
mineral waters would enjoy a greater notoriety and celebrity
than they ever did before, and their visitors be sure to return
the next season.   It would add so much to their attractions
that they could not stay away.   The grape is considered the
most healthy fruit on the Globe, and it is strange that here,
where it comes to the greatest perfection, it is utilized so little
for sanitary purposes.   As long as the American nation sub-
sists on warm biscuits and pies, washed down with tea and
coffee, intermixed with bad whiskey and brandy, so long will
it continue to be, and become more so every year, a nation
of dyspeptics.   When, in the place of these, good healthy
bread and fresh fruit of every kind becomes the daily food,
and good, sound, light wine the daily drink, we can hope for
a change for the better.   If these few lines can give an impetus, and induce but a small number to try it, they have not
been written in vain, for I know that it need but be tried to
be appreciated and followed.

CHAPTER XX.

THE VINTAGE—RAISIN MAKING.

This is a very important branch of the industry, next in importance to the wine interest.   How important it is, what
large proportions it has already assumed, and how bright the
outlook for its future, can best be seen from the circular of
Messrs. Geo. W. Meade & Co., one of the largest commission firms in that trade, and also the heaviest packers of Cal-

ifornia raisins, from which I take the liberty to quote that part relating to raisins. They say :

"We estimate the total product of 1886 at 703,000 boxes of 20 lbs each, and apportioned as follows:

Fresno district, boxes	225,000
Tulare district, boxes	8,000
Riverside district, boxes	185,000
Orange and Santa Ana district, boxes	160,000
San Diego district, boxes	25,000
San Bernardino Co., outside of Riverside, boxes	10,000
Yolo and Solano	75,000
Scattering, Yuba, Butte, Sacramento, etc	15,000
Total, boxes	703,000

" It is with a great deal of satisfaction that we approach the subject of California raisins for the year 1886. For many seasons past it has simply been up-hill work to introduce our raisins and to convince the trade generally that California could produce a fruit equal to the Malaya. While the failure to do this in a measure was perhaps due to the fact that many of the packing of California raisins were of poor quality, it is nevertheless also true, that a prejudice existed in the minds of the Eastern jobbers to that extent that they persistently set their faces against a California raisin. Notwithstanding these discouragements California has kept steadily at work improving the quality as well as the style of packing, year by year, until the outturn of 1886, on many brands, at least, equals if it does not exceed in quality the very best Spanish fruit.

" Next year we anticipate that further improvements, both in packing and labeling, will be made, so that nothing will be left that can be desired on that score.

" The labor in Spain for raisin packing is very cheap, running from 15 to 30 cents per day, while in California the same work is paid with from $1.00 to $1.25 per day. To

counteract this great discrepancy, therefore, it has become
incumbent upon the ingenuity of Californians to devise and
create machinery which would not only quicken the packing
of raisins, but would at the same time reduce the cost. The
result of this is that in California machinery is about to be
used and run by steam power, for the steming, grading, fac-
ing, and packing in the boxes of the fruit as it is received
from the grader. Machinery of this kind is almost human in
its action, but is calculated to expedite and cheapen packing,
and in a short time the cheap labor of Spain will be entirely
counteracted by the ingenuity of the California Yankee.

" All over the State this year there has been a great improvement in packing, and many of the brands produced
here rank equal or superior to the best Malaga fruit. A
proof of this fact is that such markets as New York, Boston,
Philadelphia and other large Eastern cities have taken the
finest brands of our raisins at prices equal to, if not superior
to, the Spanish goods. Different sections of the State naturally claim the best raisins, but very much depends on the
care taken in packing, sweating and properly preparing them
for market. Nearly all the various raisin sections of California can and do produce excellent raisins, and there are many
as yet undeveloped sections which can produce raisins equal
to any yet turned out. As a general proposition, it will pay
all producers to sell their fruit in the sweat boxes to some regular and reliable packer, who will maintain standard grades
from their section of the country. The policy now in use to
some extent, of small producers packing on their own account,
only produces irregular and uneven grades, and is not calculated to lead to any permanent benefit of the industry. We
can repeat our suggestions of last year, that some different
branding should be used for California raisins. As it is now,
we simply imitate the Spanish brands, when something distinctly Californian should be used.

"The trade for California raisins of good brands throughout the United States has never been as good as this year,
and at fair prices. They have been introduced where they
have never been known before, and it is now only a question
of time when, with care in packing and grading, we will entirely drive the imported raisins from America."

In a letter just received, the same firm puts the actual quantity of last year's crop at 750,000 boxes, considerable above
their estimate; and estimate this season's pack at 1,000,000
boxes, one fourth more than last year.

The Cailfornia process of raisin curing differs from the
European, and is thus described by Mr. N. B. Blowers, one
of the veterans in that culture:

The grape should be allowed to remain on the vine until
quite ripe, showing a yellowish or golden color, and being
more translucent than when too green. Then they should be
carefully picked, and placed upon a drying tray, usually two
by three feet in size, and exposed with an inclination towards
the sun, in some convenient place, generally between the rows
in the vineyard. After being sufficiently exposed to become
about half dried they are turned once in this manner, viz.:
two workmen taking an empty tray, place it upon a full one,
holding them together firmly, and with a swinging motion
turn them over, and replace the now turned grapes in their
former position. The turning should be done in the morning,
before the dew is quite off the grapes ; then, when the grapes
have become so dry as to loose their ashy appearance, some
being a little too green, and others quite dry enough, they are,
after removing those that are entirely too green, slid from the
tray into large sweat boxes, having a thick sheet of paper between every twenty-five or thirty pounds of raisins, then they
are removed to the storeroom where they should remain two
weeks or more. When ready to pack it will be found that
the too moist ones have parted with their surplus moisture,

which has been absorbed by the stems and drier raisins. The stems are now tough, and the raisins soft and ready to pack. They are carefully placed in frames made of iron or steel. The large and fair ones being carefully placed in the bottom of the frames, the surplus stems and berries cut away ; then the average raisins are arranged in and weighed, placing five pounds in each frame and pressed, but not enough to break the skin. They are then passed to an inspector, who examines the exposed side of the raisins, removing all imperfect ones, then placing the wrapper paper on the frame, holds it in place with a wooden or steel plate, turns it bottom up, drops the left end into the box, slides the plate quickly from under the plate and it drops into the box, then pressing slightly upon the movable bottom of the frame, the frame is removed. The bottom of the frame is then pressed more firmly, to cause the raisins to fill the space formerly occupied by the sides and ends of the frame, then it is removed and the face of the latter is exposed, all imperfect berries or too wet ones are removed, and all vacancies or hollows filled with large, loose raisins. The label of the proprietor is then placed on the face; the ends of the wrapper, and then the sides are folded over, the box cover nailed on, and they are ready for market.

The favorite varieties for raisins are the Muscatelle Gordo Blanco, or Muscat of Alexandria, while a very fine seedless raisin, but much smaller, is made from the Sultana. The Corrinths, white and black, so far have not proven sufficiently successful here to warrant their extensive culture.

There is a difference of opinion yet among raisin men, whether irrigation is absolutely necessary or not. The irrigationists claim that the berries are larger, more uniform and showy, while the other side claims finer flavor and more delicate bloom. Not being a raisin grower myself, I am not able to decide which is right, but most of the raisins so far have been produced under a system of irrigation.

During the last year or so, raisin growers have had a net return of from 120 to 200 dollars per acre, and found the industry very profitable, while, when they sell their crop in the sweat boxes, as is the general custom now, they have comparatively light expenses, as compared to the wine maker. The future of the California raisin trade seems to be secured, and with our rainless falls we have superior facilities for drying to any country on the globe.

It is also an open question yet, whether the heat of the sun, or artificial heat in driers are to be preferred. Without going into this question further, we may rejoice in the undisputed fact, that every one *can* make raisins without artificial heat.

As a further evidence how California raisins are appreciated, I copy the following from the *New York Mail and Express:*

CALIFORNIA RAISINS AHEAD.

Competing Successfully With the Spanish Fruit in New York.

The California raisin has reached such perfection that it is now able to compete successfully with the finest of the Spanish fruit. Already some of the present season's crop is in the market, and presents a handsome appearance. The loose native Muscatel is now packed by machinery, which has enabled the packers to compete with the low-priced manuel labor of Spain. The "steam power stemmer" stems and divides the fruit into three grades at the rate of 100,000 pounds a day, with the assistance of about twenty men. This rapidity of operation would surprise the Spanish packers with their primitive methods of stemming by hand. To this is added a packing machine, which packs the stemmed raisins in boxes of twenty pounds weight each. A well-known handler said this morning:

"I must acknowledge that this season the California raisins are superior to the imported Spanish fruit. The grape crop has been large and fine, the raisins better cured, and, I think, will command nearly if not quite as high a price as the foreign product. Were it not for the Interstate Commerce law, which has increased the cost of transportation from twenty to thirty-five cents a box, the native fruit could be sold cheaper. It seems strange to me that the government should thus

15

impose a tax upon the native products of the country. I am informed
that over 1,000,000 boxes, or 50,000,000 pounds, of raisins are coming
Eastward from California. Of this quantity, about one-tenth, or 2,000,-
000 pounds, will reach New York. A part of this shipment has already arrived, and more will do so next week. So fine is the fruit that
it is bought up to a great extent before it reaches here. The remainder of the Eastern shipment is dropped by the way at the principal
cities. It is my opinion that the California raisin will soon drive the
foreign out of the American market, and will finally be exported to
Europe."

CHAPTER XXI.

INDIVIDUAL ENTERPRISE.

It will be impossible, in the limited space allotted to me,
to do justice, or even mention, all the important individual efforts in grape growing, wine making and cellars, for which
our State is already justly famous ; nor would it be of any
real benefit to enumerate them all, were such a thing possible.
But a short description and mention of a few of the largest
and most important should find place, as part of a picture of
this immense industry.

The largest vineyard in this State is that of Senator Leland
Stanford, at Vina, Tehama County, on a piece of rich alluvial bottom land on the banks of the Sacramento River.
On his magnificent ranch of over 50,000 acres, an area of
about 3 miles is now planted in vines.   The oldest vineyard
is now about thirty years old and was planted by a Mr.
Gerke.   To this have been added in 1882; 1,021 acres, in
1883, 1,053 acres, in 1885, 900 acres, a total of 3,054 acres.
All this immense tract can be irrigated if thought advisable,

but so far only the young plantations have been watered the
first year, to give the cuttings a start, and there is no necessity for further irrigation. The leading varieties are Zinfandel, Trousseau, Blaue Elbe, Burger, Sultana, Mission, Malvasia, Orleans Riesling; although Sauvignon, Franken and
Johannisberg Riesling, Chas. Fontainebleau, Black Burgundy,
Mataro and others are planted on a smaller scale. The vines
are of course, not in full bearing yet, but they produced last
year about 2,500 tons, while the crop this year was estimated
at 5,000 tons, though it may have fallen somewhat short, as
in most of the vineyards in the State. On one corner of these
immense vineyards the wineries, distillery, etc., are located,
covering about 4 acres of ground. They are: 1. The old
cellar, two stories high, 105x157 feet, capacity 500,000 gallons. 2. The new fermenting house 101x169 feet, two
stories high, capacity 500,000 gallons. 3. The new cellar,
266x297 feet, capacity 1,700,000 gallons. 4. Distillery,
32x95 feet. 5. Brandy warehouse, 34x100 feet. There is
the old Gerke cellar additional, capacity 500,000 gallons,
and a new bonded warehouse, 34x100 feet.

The machinery is all run by steam, of Heald's newest improved pattern of crushers, stemmers and hydraulic presses,
4 crushers, 4 presses, and can work up 400 tons of grapes per
day. The casks are all oak, of a capacity of 2000 gallons each,
the fermenting tanks of redwood 10 feet wide, with a capacity
of 2,400 gallons each. Capt. H. W. McIntyre, the superintendent of the cellars, is one of the best architects in the State,
and has spared no pains to make the buildings and machinery
as complete and practical as possible. Mr. Smith is Farm
Superintendent, Mr. Shackleford Civil Engineer. Most of
the new buildings were put up in the course of this summer,
and are of brick. During the vintage and erecting the buildings a force of about 1,200 men were employed on the place.
All the cellars are lighted by electricity, so that a force can be

kept working night and day, if necessary. When Gov. Stanford first conceived the idea of planting such a large vineyard, he declared his object to be to furnish cheap, wholesome wine to the million, so that every laborer could drink it. In a few years the vineyard bids fair to fulfill it. I doubt whether very fine, light dry wines can be made there; the tendency of soil and climate would seem to be more favorable to sweet, heavy wines, and the manufacture of brandy. But if good, dry wines can be made there, (and this season's operations seem to prove it), the present management will make them, and with all the facilities for fermenting and regulating the temperature I have no doubt that sound wines at least can be produced, and the object of Gov. Stanford be attained.

The next largest vineyard in the State is the Nadeau vineyard in Los Angeles Co., but my information about it is not as complete as I could wish, as the manager, Mr. Eggleston, did not respond to my enquiries, and the notes I have were only furnished me by the courtesy of the officers of the State Viticultural Commission. These show 2,401 acres of vineyard, of which 1,400 acres are Mission, 466 acres Zinfandel, 466 acres Riesling, balance mixed varieties. There were on hand some time ago, 50,000 gallons Brandy, 4,800 gallons Angelica, 5,000 gallons Port, how much dry wine, I am unable to tell. This would seem to indicate a tendency for the manufacture of sweet wines and brandy. The vineyard is still young, and has not attained its full bearing capacity.

The next largest, and the most important perhaps as a factor to solve many of the problems of Viticulture in this State, is that of the Natoma Company at Folsom, near Sacramento. I visited the vineyards in 1884, together with its then manager, Mr. Horatio P. Livermore, to whom I am indebted for most of the information I give now; and I considered it then the most promising large enterprise on this coast. The subdivision into 400 acre tracks, each with its foreman and separate working force, with buildings in the center, affording convenient quarters for all the men necessary, as well as for the work animals, barns, tools and tool sheds, struck me as the only possible way to work large vineyards successfully; and I could not help but admire the organisatory talent of Mr. Livermore, which was evident in every direction. Everything seemed to go like clockwork; the young vineyards were in a high state of cultivation, showing extraordinary growth and even some fruit; and the many new varities tested then, would have been of incalculable benefit to our young industry, if the plans of Mr. Livermore had been adhered to. I think it a public calamity that he was forced to abandon a task he had so well begun; although the vineyards may have fully come up to the expectations of the present managers in a pecuniary point of view, the benefits which would have accrued to the grape growing public at large, were in a great measure lost sight of. The experiments with new varieties, imported by Mr. Livermore, and which such a company could have carried out better than individuals, were virtually abandoned; and although the varieties were in a measure distributed over the State, and thus not altogether lost, the testing at Folsom or by the Company was abandoned. It is because I think this enterprise so important in its results to the wine growing interests of the State, that I have given it so much space in these columns. The soil at Natoma upon which most of the vineyards are located, is a red, volcanic soil, intermixed with stones and pebbles, naturally well drained, and would seem to indicate high quality of wines. But the climate is very warm, somewhat malarious, and it may be more adapted to the production of heavy sweet wines and brandy, than to the finer light dry wines.

They are situated in Sacramento County, California, about sixteen miles northeast of the city of Sacramento, the capital of the State. The Sacramento Valley Railroad runs for three miles

through the vineyards, affording three switches for shipping
purposes, the central one of which is a regular station with
depot, post office and express office.   It is called " Natoma "
and here are situated the company's business offices, so that
no part of the vineyard is over one mile distant from a railroad, which affords connection with the entire railroad system of the State, also with the transcontinental railroad systems.   The vineyard proper forms the southerly portion of
the " Rancho Rio de los Americanos " an original Mexican
grant of about 9,000 acres, duly patented by the United
States Government, which the Natoma Company purchased
not many years ago and still owns almost in its entirety.   It
lies on the east bank of the American River for a distance of
seven miles and includes within its borders the town of Folsom, for many years of the first importance as a placer mining centre.

The vineyard plantation nowhere comes down to the
border of the American River, but is confined to the benches
or slopes that run back, and merge into, the rolling foot-hills.
Its elevation ranges variously throughout its extent from 150
to 300 feet above the sea level, and it is distant from the
Pacific Ocean about 125 miles in a direct line (westward)
shut off therefrom by the Coast Range chain of mountains,
which average from 4,000 to 6,000 feet of elevation ; while
from the Sierra Nevada range of mountains, which dominate
all California, and here have an elevation of 8,000 to 12,000
feet, it is distant about 40 miles.   Thus the climate is essentially an inland one, sheltered from all sea winds and fogs,
scarcely ever exposing the vineyard to frost damages, but
visiting it, in the vintage season, with extreme heat (sometimes as high as 105° F in the shade) which powerfully stimulates the growth of the vines and the development of sacharine matter in the grapes, thus making the excess of sugar in
the grapes a thing to be guarded against in the making of dry

wines, and indicating its best specialty as perhaps in the direction of ports, sherries and other full-bodied types of wines.

The soil is a red loam, from three to twelve feet in depth, underlain by gravel, and it with cobbles, so that it may be said to be exceptionally well drained, an important feature, for the winter rainfall, doubtless owing to its proximity to the Sierras, is very considerable, sometimes even excessive.

The vineyard is drained by Alder Creek and Buffalo Creek, two water courses which run across it from east to west and carry its surplus waters into the American river, which has so heavy a fall as to rapidly dispose of all floody waters. The lands were heavily timbered with oak trees, many of them of great size and very deeply rooted, requiring heavy expenditure for clearance in preparation for vineyard planting.

The first vineyard planting of the company was in the year 1876-7, when about 110 acres were put out—seventy acres to Muscat of Alexandria and forty acres to Flaming Tokay. The "Muscat" was planted with reference to raisin making, and was the earliest considerable planting of that variety for that purpose in the State—it may be stated here that it proved flourishing and bore vigorously, but after trying for three years the experiment of raisin making, the company concluded that there was too great liability to early fall rains to permit of reliable *field curing* of raisins, so they abandoned it as a raisin vineyard, and have since sold its product to the canners and to the Eastern shippers, finding the latter always willing to give a preference to the Natoma product because of size, excellent flavor, and special ability to stand distant transportation. This initial plantation was followed in 1879-80 with seventy acres more; viz., fourty-five acres Zinfandel and twenty-five acres more of Flaming Tokay.

In 1880-81 were put out fifty acres more — of which thirty acres Tokay, ten Emperor, five Seedless Sultana, five Black Ferrara.

In 1881-2 were planted about seventy acres, mostly Flaming Tokay and Zinfandel—this brought the entire plant up to 300 acres, all of which except the Zinfandel was of shipping varieties of grapes, and consequently this part of the company's vineyard has always been known as the "Shipping Vineyard," and its products have always commanded a ready sale for shipment to the Eastern markets. It lies directly on the railroad, at no part distant more than half a mile therefrom, with a very extensive packing house, so that two entire cars can be loaded at the same time, and a force of 200 hands can be accommodated at once, picking, selecting, and packing the grapes, which are placed without any jolting of wagon transportation, directly on the cars, and are the same evening over the Sierras on their way to Eastern markets, at least one day ahead of the coast counties.

The grapes produced in this vineyard are large, firm berries, full clusters, and the Tokays are of extraordinary size of bunch, brilliancy of color, and durability to stand transportation. For the last three years shippers have freely paid fifty-five to sixty-five dollars per ton for them on board cars, they furnishing packages.

That shippers are justified in giving preference prices for this pack of grapes seems to be established by the Eastern market sales quotations, where the Natoma brand uniformly brings a higher price than other packs (E G Chicago, Sept. 5th, '87 quotes Natoma Tokays sold average $2.85 per crates, while other brands were quoted $2@.$2.25 per crate.) All this shipping vineyard is under a complete system of irrigation from the company's own canal, which carries a large supply of water throughout its entire track and the adjacent country, from Salmon Falls on the South Fork of the American River, about twenty miles distant from the vineyard.

Very great judgment and *moderation* have to be practiced in applying this irrigation, but it is the company's experience,

that, with such care, a timely drink or two to the vines, in
the growing season, increases the crop, strengthens and am-
plifies the foliage (so important in prevention of sunburn to
clusters of fruit), enlarges the berries, fills out the bunches
and generally promotes the quality as well as the quantity
of the yield.

Many acres of this shipping vineyard have, under this system, yielded seven and eight tons to the acre of shipping
grapes, but it is safe to say that, taking the average of years
and of acres, at least four (4) tons to the acre, of selected
grapes, in crates on the cars, may be counted upon from the
vineyard for each year, and since the quantity of culls and
trimmings, not packed, that go to the wine house and distillery, go very far towards paying all the expenses, it follows
that this remarkably favored vineyard realizes upwards of
$200 per acre per year. The Tokays are grown well up from
the ground, are staked and pruned with medium long canes,
special attention being paid, in the season of growth, to
nipping back the long shoots, so as to make denser the foliage,
and thus protect the clusters of fruit from that worst foe of
the Tokay grape, viz : "sunburn."

In the season of 1882-83 it was realized that so complete a
demonstration of the fitness of Natoma soil and climate to
the growth of the grape had been made as to justify a much
larger utilization of the company's lands for vineyard, and,
accordingly, the company directed the then manager (one of
the principal stockholders as well), Horatio P. Livermore, to
proceed with the immediate planting of 1000 acres more of
vineyard.

Commencing work on November 10th, 1882, the land was
cleared of its timber, thoroughly subsoiled, and planted by
March 23d, 1883, work which illustrates strikingly the favorable character of the California winter climate, for nowhere
else would it have been possible, because of probable inter-

ruptions by stormy weather. This season's plant amounted to 965 acres, as follows :

Verdal, 20 acres ; Malaga, Blues White, 5 acres ; Black Ferrara, 10 acres ; which, being all shipping varieties, were planted in extension of the *shipping* vineyard ; and the following exclusively Wine varieties, viz : Lenoir, 10 acres ; Zinfandel, 150 acres ; Feher Zagos, 15 acres ; Meunier, 15 acres ; Crabb's Black Burgundy, 50 acres ; Chalosse, 10 acres; Columbar, 50 acres ; Grenache, 60 acres ; Chauché Noir, 65 acres ; Sauvignon Verte, 25 acres ; Mataro, 75 acres ; Moselle Riesling, 20 acres ; Orleans Riesling, 5 acres ; Franken Riesling, 20 acres ; Wests Prolific, 5 acres ; Seedless Sultana, 10 acres ; Piquepoule Gris, 3 acres ; Frontignan, 12 acres ; Trousseau, 50 acres ; Berger, 85 acres ; Malvoisie, 25 acres ; Carignane, 70 acres ; Caberenet Malbec, 20 acres ; Charbonneau, 30 acres ; Folle Blanche, 50 acres.

This plant was made entirely of cuttings, and sufficient more cuttings, of the same varieties, were placed in nursery for rooting, to similarily plant another 1000 acres in the succeeding year, which the Company directed the manager to prepare for, the intention being to carry the vineyard up to 3000 acres.

Small experimental plantations were also made, for test purposes, of some of the American resistant varieties, Herbemont, Rupestris, Cynthiana and Norton's Virginia ; and extensive propagating of Lenoir was undertaken for future planting.

The Company having on a tract of its lands, adjoining its Orchards, three miles from the vineyard, and near the town of Folsom, a plantation of thrifty Mission grape vines, upwards of ten years old, it was decided to graft these into the choicest wine varieties which could be imported. Accordingly, at very considerable expense, there were imported from France, Spain, Italy, and Portugal, the cuttings of the following varieties, and

grafted into upwards of twenty thousand old Mission vines;
Carbernet Sauvignon, Carbernet Franc, Merlot, Verdot, Mal-
bec, Semillion Blanc, Sauvignon Blanc, Muscadelle de Bor-
delais, Aramon, Petit Bouschet, Mourastel, Cinsaut, Beclan,
Poulsard, Serine, Mondeuse, Clairette Rouge, Pécoui Touar,
Clairette Blanche, Ugni Blanc, Rousanne, Marsanne, Tannat,
Petite Syrah, Malmsey Madeira, Tinta Madeira, Verdellho,
Boal, Muscatel Madeira, Pedro Ximenes, Palomino, Mantuo
Castellano, Veba, Péruno, Mantuo de Pilas, Bastardo,
Mourisco Preto, Tinta Coa, Morete, Mourisco Blanco,
Tinta Amarella, Touriga, Bokador, Yellow Mosler, Pever-
ella, Rothgipler, Rhulander Grey, Slankamenka, Yellow
Silk Grape, Steinschiller, Green Sylvaner, Spicy Tramin-
er, Green Veltliner, White Vernaccio, Waelschriesling,
Zierfandler, Affenthaler, Kadarka, Lagrein, St. Laurent,
Marzemino, Portugieser, Refosco, Spanna, Barbera, Terol-
dego, Wildbacher, Malvasia Bianca, Moscato Rosa, Rosara,
Aleatico, San Giovetto, San Columbano, Trebbiano, Cana-
jola Nero, Canajola Bianco.

Satisfactory success was attained with the most of these
varieties, and thus was established a store house of viticultural wealth for the State, which subsequent vineyard planters
have largely and profitably availed themselves of.

It was fortunate for the State that this work could be
undertaken by a corporation wherein those interested were
few in number and had ample means, and whose property
was so favorable, in all respects, to such experimental test
work, and great benefits will undoubtedly result to the State
of California therefrom, though, since Mr. Livermore's retire-
ment from the active managership of the Natoma property
(which took place in 1885) not all his wise and public spirited plans have been carried out by his successors.

Continuing the plantation in the years 1883-4, a decidedly
unfavorable season was encountered, owing to the protracted

delay in the usual winter rains, rendering it impossible to commence, seasonably, the preparation of the ground, and consequently very much curtailing the planting season, so that there was planted but 600 acres of the 1,000 acres projected, the varieties being as follows :

Zinfandel, 100 acres ; Columbar, 16 acres ; Trousseau, 80 acres ; Purple Damascus (in shipping vineyard), 5 acres ; Mataro, 100 acres ; Chauché Noir, 36 acres ; Chalosse, 24 acres ; Grenache, 65 acres; Berger, 30 acres; Folle Blanche, 40 acres ; Riparia, 7 acres ; Pedro Ximenes, 4 acres ; Petit Bouschet, 3 acres; Mantuo dé Pilas, 2 acres ; Mondeuse, 3 acres ; Bastardo, 2 acres ; Palomino, 10 acres; Tannat, 6 acres ; Roussanne, 5 acres ; Muscadelle de Bordelais, 5 acres ; Petite Syrah, 6 acres ; Carbernet Franc, 3 acres ; Verdelho, 4 acres ; Tinta Madeira, 4 acres; Malmsey Madeira, 4 acres ; Sauvignon Blanc, 3 acres.

Fractional experimental blocks composed of the varieties following :

Côt a que Rouge, Côt a que verte, Carbernet Sauvignon, Semillion Blanc, Pineau D'Aunis, Tinta Madeira, Muscatel Spanish ; 33 acres in all.

There was thus presented an entire planting of upwards of 1800 acres, of which 200 acres, approximately, were shipping and canning varieties, and did not, consequently, interest the Wine House, except so far as their cullings, in packing for shipment, went to the distillery. Of the strictly wine making acreage the prominent factors will be seen to be the following varieties: Zinfandel, 350 acres; Crabb's Black Burgundy, 50 acres ; Chalosse, 34 acres ; Columbar, 66 acres ; Grenache, 125 acres; Chauché Noir, 100 acres; Sauvignon Verte, 25; Mataro, 175 acres; Trousseau, 130 acres; Berger, 110 acres; Carignane, 70 acres; Malbeck, 20 acres; Charbonneau, 30 acres; Folle Blanche, 90 acres; Meunier, 15 acres;

and these varieties, therefore, are what the company will have for the present to build the reputation of their wine upon.

In the earlier days of their yield it was not deemed advisable to make wine of their product, and it was sold to Messrs. Kohler & Van Bergen, of Sacramento, and by them made into wine with results reported as most satisfactory.

In this year, however, (1887) the company proposes to make its own wine and to that end has erected the first section of its wine house, calculating upon a capacity of three hundred thousand gallons, to be increased in succeeding years as the increased yield of the vineyard calls for larger accommodation.

All these varieties are reported as showing a good crop this year, considering the age of the vines, and if they behave as well in the wine house as they have in the vineyard, some very interesting results may be expected, as each kind will be made separately.

The wine house is situated, not on the railroad track (as is generally preferred when there is a railroad) but on a bluff nearly in the center of the wine vineyard, half a mile from the railroad.

It is one story high, covering ground space 96x130 feet, doubled walled, leaving central air space, is of wood, ceiled inside with tongued and grooved lumber, also similarly ceiled above to rafters, so as to make practically a double roof. Eight ventilators regulate the temperature, which, it is claimed, has been found easy to control, so as to keep a temperature of 75° F in the wine house when upwards of 100° in the shade outside. The roof is divided into two gables and a two (2) story tower is built at the end, to carry elevator and crusher, which is Heald's latest improved, operated by steam engine, and handles forty-five or fifty tons per day of ten hours.

Under said tower on front line of the wine house runs the driveway for loaded wagons to elevator and crusher. Hydraulic press are also conveniently located under the tower. The fermenting room holds forty fermenting tanks, five feet deep by nine feet across, capacity 2000 gallons each. There are ninety-six storage tanks, eight feet high by six feet across; capacity, each 3000 gallons.

No provision is yet made for permanent storage and maturing wine for quality, and it may be that delicate wines, which it is desired to mature, will be carried in some warehouse located on San Francisco Bay, at a point suitable for shipping and for maturing wines.

A duplicate of this wine house will be built, in extension of the present one, next year, to accommodate that year's increase of product, and similar additions will be made as product increases.

Conveniently adjacent to the wine house, in a building of corrugated iron, is a brandy distillery, of Sanders & Co.'s most approved make, rated to produce, from sound wine, 804 gallons brandy each, in twenty-four hours, from sour wine, 574 gallons, from pomace, 251 gallons. Water tank houses are attached to both wine houses and distillery, and water pipes are carried through both buildings for use and fire protection.

The vineyard is most liberally planned and laid out. Each block is marked with its variety at each corner, an exterior avenue all around it twenty-four feet wide, and a similar one through the middle, parallel with the railroad, allow four horse teams to be readily turned, when plowing, cultivating, &c. The vines are carefully staked with redwood stakes, the whole track is very securely fenced with a rabbit-proof fence. For convenience of working it is divided into sections of 400 acres, each section having its foreman's house, barn for work stock, &c. The working is centralized at the headquarters house, near wine house, where reside the superintendent and accountant with their families. The buildings are all first class, and admirably adapted for their purposes, but are plain and inexpensive.

Altogether it may safely be said that the Natoma Vineyards are a most striking illustration of the rapid advance of the viticultural interests in the State of California, and are a lasting credit to the judgment and ability of Horatio P. Livermore, who planned and executed them and who, had he continued to manage them, would have undoubtedly worked out many results from the extensive importations of foreign varieties there grouped together, which would have proved of the greatest benefit to the State at large in its viticultural interests.

It is to be regretted in this regard, that the management succeeding Mr. Livermore, has failed to carry out his plans, has neglected to prosecute many of the interesting experiments, so wisely instituted at Natoma, and has preferred to run the property solely for immediate commercial results. It is believed that thus have been lost many opportunities of great future enhancement of the value of their own property, and of concurrent inestimable benefit to the viticultural interests of California; for never in California were there grouped in any one vineyard, so many promising features as at Natoma, and no where could they have been so readily and profitably worked to a fruitage, had the policy which originated the vineyard been maintained.

One of the important enterprises of this kind are the Sunny Slope vineyards, formerly owned by Mr. L. J. Rose, near San Gabriel, California, but lately sold to an English syndicate. Mr. Rose, Viticultural Commissioner for the Los Angeles districts, is one of the pioneers of grape culture in the State, and some of his vineyards are sixty years old, being among the oldest in the State, and yet in a flourishing condition. In 1886, Mr. Rose made, at his extensive wineries and distillery 225,000 gallons of wine, of which 100,000 were red wines,

125,000 white wine, and 60,000 gallons of brandy. I have no later dates. His leading varieties are Trousseau, Mission, Mataro for red, with some Crabbs Burgundy (Refosco) Blaue Elbe, Burger and Sauvignon or Pedro Ximenes for white.

The San Gabriel Wine Company, J. de Barth Shorb, President and Manager, have also a very large establishment, but as I applied to Mr. Shorb for information, and received only a very curt reply, that he had no time to give it, I am unable to give particulars.

A very important enterprise is that of Mr. Juan Gallegos, at Mission San Jose, Alameda county, also lately sold to a company; which, however, purchased only the winery and water right, casks, tanks, and 250,000 gallons of wine, at a valuation of $300,000 ; Mr. Gallegos reserving the vineyard of 550 acres, but agrees to sell the grapes to the Company. The winery has a capacity for storing one million five hundred thousand gallons on first and second floors, while the third story has a capacity for fermenting one million gallons. The wine house is built of brick and stone, 240x110 feet and contain two crushers, two elevators, one hydraulic press, and all the necessary apparatus, casks, tanks, etc., besides the distillery 35x85, and a frame building for fermenting purposes, cooper shop, etc., with abundance of water and water right. The vineyard is 550 acres, containing the following varieties.

RED.

Zinfandel 435 acres, Mataro 34 acres, Cabernet Sauvignon 24 acres, Burgundy 16 acres, Trousseau 6 acres, Tannat 5 acres, Mondeuse 4 acres, Petit Bouschet 4 acres. Total, 528 acres.

WHITE.

Green Riesling 11 acres, Sauvignon Vert (Pedro Ximenes) 8 acres, Clairette Blanche 3 acres. Total, 22 acres.

While admiring the enterprise of Mr. Gallegos, in building

so large a winery, and planting so large a vineyard, I cannot help but warn my readers against such a selection of varieties. If they will examine, they will see that there are but twenty-two acres of white wine varieties, ninety-three acres of other red wine varieties, and 435 acres of Zinfandel. Only imagine over three-fourths of the whole crop of one variety, ripening at the same time, and as difficult a variety also as the Zinfandel. I would not like to be wine maker there, and think this is one of the examples " how not to do it " in the selection of varieties.

While these are, perhaps, the largest enterprises of their kind in the State, there are hundreds, fully large enough, and which were built up from small beginnings by the owners themselves. Among these, Napa Valley may claim a prominent part. Among those who have their own vineyards and mainly work up their own grapes are Mr. G. Groezinger, at Yountville, who has a vineyard of 450 acres, and made about 700,000 gallons of wine last year; Mr. H. W. Crabb at Oakville, vineyard of 350 acres, and a vintage of half a million gallons last year. Mr. Crabb is a living example of pluck and enterprise. Commencing very small and in very unfavorable times, he has now one of the most flourishing vineyards in the State, his winery and distillery are a small village by themselves; his wines have a reputation not confined to this coast, but widely spread over the Eastern States. He has a collection of over 300 varieties of grapes, fruiting and growing on his own premises, and is so familiar with them that he is unquestionably the best authority on grapes in the State. Public spirited in the best sense of the word, all the information he has gained is at the service of everyone, and he is always prominent wherever the interests of the grape growers can be served best. Such men are the benefactors of their race and of their calling, and it affords me pleasure to acknowledge the debt of gratitude we owe them. But I think I also

16

repeat but the wish of his many friends, that he should not
strain his powers too much, as he must necessarily do, conducting all his business nearly alone. Such lives are too valuable to be shortened by overwork. Another of our pioneers
is Mr. Chas. Krug, and although at present rather "under a
cloud," brought about by his over-sanguine temperament, he
has done a great deal to advance viticulture to its present
prominence in Napa County, and his energy and pluck will
soon overcome all temporary obstacles; Napa Valley, and especially St. Helena, owes him much of its present prosperity
and prominence. Mr. Wm. Scheffler, of Edge Hill Vineyard, Beringer Bros., Gustave Niebaum, M. M. Estee, are
among the largest producers of the valley, and Messrs. Ber-
ringer Bros., whose cellars are excavated from the solid rock,
and Capt. Niebaum's at Rutherford, designed by Capt. Mc-
Intyre, strive for the palm of the best cellars in the valley ;
which is not a small honor when we consider that there are
over two hundred wine cellars in all within a distance of about
twenty-five miles, with a capacity of storing 6,000,000 gallons. The wine makers who purchase grapes and manufacture them into wine are mostly located at Napa, and among
these the Uncle Sam Wine Cellars, formerly C. Anduran &
Co., now, since the death of Mr. Anduran, Carpy & Co. is
the largest. They manufacture mostly clarets for the New
Orleans and Southern trade, have a capacity of nearly a million gallons, and will make about 6,000,000 gallons this year.
They not only purchase grapes from Napa County, but also
from Solano, Yolo, and Santa Clara. The average price has
been fifteen dollars per ton, delivered at Napa, for Zinfandel
and other good varieties; from ten to twelve dollars per ton
for Mission and Malvasia. G. Migliavacca and sons have also enlarged their wineries to about 350,000 capacity, and the
Napa Valley Wine Co., Mr. Priber, manager, may work up
about the same quantity. Mr. Mathews, of the Lisbon

winery, manufactures mostly Sherry and Port, and his winery, to a large extent built by himself, is a striking evidence of individual skill and enterprise.

Sonoma county is the friendly rival of Napa in the production of fine wines and large enterprises. Mr. J. de Turk, the commissioner for that district and also one of the largest wine makers, reports the crop of 1886 as 3,500,000 gallons made from 25,000 tons of grapes. Mr. de Turk reports this season's crop to be about 33 per cent. short of last year, owing to frosts, coulure, and dry weather during vintage. Among its largest producers and dealers are J. de Turk, E. H. Sheppard, Kohler & Frohling, Chauvet, Aguillon at Sonoma, and especially the venerable pioneers of grape culture, Mr. Craig, J. Gundlach, Julius Dresel. To these last especially belongs the honor of making a reputation for choice white wines for the State, as well as taking the initiatory steps of combating the phylloxera by the introduction of American vines. I venture to say that there are not finer white wines to be found in the State than at Gundlach's Rhine farm at Sonoma, or the adjoining cellars of J. Dresel & Co., nor can a more striking illustration of the complete resistance of the Riparia be found than in their vineyards, devastated by the insect ten years ago, planted with Riparias in 1880, 1881-82, and since grafted with choice varieties. They are the pioneers in this, and have done as much for the permanent establishment of grape culture, even more, than many a one who has planted a thousand acre vineyard.

Capt. J. H. Drummond, of Glen Ellen, has done a great work by importing the choicest varieties of vines, trying different methods of training them, and by his choice wines, which have been favorably noticed at every exposition, demonstrating the fact that California is qualified to rank with any part of Europe in the production of choice wines. Mrs. Kate F. Warfield and Mrs. Hood have demonstrated that

women are as able to manage vineyards and wine cellars, as
the majority of men, and altogether old Sonoma has a good
record to show of individual effort and pluck.

Santa Clara County can also show many instances of individual enterprise. Foremost among them is our lamented
friend Chas. Le Franc, of whose untimely death I was informed but a few days ago. I will remember his kindly face
and the royal welcome at his winery in 1881, which
greeted the then stranger to this coast, and his fine wines, of
which we partook at his wine cellars. Both he and Gen.
Naglee have "gone to their fathers," but ought not to be
forgotten when the muster roll of the Pioneers in Grape Culture is called. Santa Clara has many large vineyards, Capt.
Merithew, John T. Doyle, J. B. Portal, J. Pfeffer, Paul O.
Burnsbe and others, and claims to make as good red wines
as any part of the State. From here, many of the choice
Bordeaux varieties were first distributed, and the greater
part of their plantations were of that class.

In Fresno County, the Barton Estate Company now own
the vineyard planted by Mr. Barton, although he still retains a
large part of the stock, and is made the managing director for
three years. This vineyard contains about 500 acres, but
it is the intention to purchase and plant 320 acres more.
Last season's crop was about 270,000 gallons, and about
300,000 are expected this year. They have fine cellars and
fermenting houses; and everything is planned on a most extensive scale. Captain Eisen has the oldest vineyard there,
of some 400 acres, and extensive cellars. Mr. Eggers has
also a large winery, and Lachman & Jacoby are interested in
another. Fresno produces very heavy wines, and would
seem to indicate a special adaptation to Angelica, Port and
Sherry, while I doubt that it will ever produce fine, dry,
light wines. It also excels in raisins, and in these two
branches of the grape industry it has a great future. It

would lead too far if I enumerated all the individual enter-
terprises in the State. There are so many that this would be
an endless task. We have enterprising wine men in all sections of the State, and it would seem next to impossible to
enumerate them all.

But it would not be fair to forget our dealers, who have
done so much to find and open a market for our wines.
Among those who were among the first growers of grapes, as
well as dealing in wines, I will name our departed friend
Chas. Kohler, who started the oldest winery in Los Angeles,
and was one of the first to enter the markets. J. Gundlach,
Julius Dresel, and Chas. Krug all entered the market at an
early day, and also contributed greatly, by their importation
of choice foreign varieties, to raise the quality of our product.

Nor should Col. Agaston Haraszthy and his son Arpad, be
forgotten. Col. Haraszthy imported perhaps the largest collection of foreign vines into the State at an early day, and the
industry is greatly indebted to his efforts. Mr. Crabb has
opened quite a large Eastern trade, and many of our large
growers are following these examples. Dreyfuss & Co. have
large vineyards and wineries near Anaheim, and also one of
the largest wine houses in San Francisco. G. Eschelbach,
near Santa Ana, is a large grower and dealer, so are J. de
Turk at Santa Rosa, J. L. Rose, and the San Gabriel
Wine Co.

Of those who have only dealt in wines, without producing
them, perhaps S. Lachman & Co. of San Francisco, are the
most prominent. Their new wine house on Brannan Street
is a monster institution, will easily store a million gallons,
and is a lasting monument to its builder, Mr. S. Lachman.
It combines great durability, immense storage capacity, and
ventilation with convenience for working and handling all
kinds of wine, which can hardly be excelled anywhere.

I have tried to give a faint outline of the individual enterprise in California. I am aware that I cannot do justice to all, that many were omitted who well deserve a place. But my readers must take the will for the deed, and while I have tried to give a brief sketch of some embarked in this great industry, it was next to impossible to remember all and do them justice. This would require a separate volume, and an abler, though perhaps no more willing pen than mine; and it would be an interesting task indeed to write a history of the Pioneers of the Wine Industry, which I hope some of my "brethren of the quill" will undertake, before their "deeds of peace" have become obsolete, and their memory is blotted out by the hand of time.

CHAPTER XXII.

CO-OPERATION IN VITICULTURE.

While I am aware that individual enterprise has already worked wonders in this, as in every branch of industry on this Coast, and no one can be more willing and ready to appreciate its efforts, yet I believe that still more could be accomplished by co-operation of the grape growers; throughout the State as well, as by combined efforts in each community and district.

Our Viticultural societies, State as well as in each District, have done a great deal of good already, by their meetings and discussion to diffuse knowledge by interchange of opinion, comparing samples of wine and grapes, etc. They should be attended by every grape grower ; it will be a change from his daily routine, and he will return to it with knowledge gained, new ideas awakened, feeling more able to cope with his task, and encouraged by the success and example of others. We should meet in a friendly feeling of rivalry, with tolerance of the views of others, and ready to impart what we know, receiving the knowledge of others in return. We have two important institutions in common, the State University and its Viticultural Department, and the State Board of Viticulture. Both have already given us varied information, and are continuing to do so. Let us strengthen their hands by ready and willing co-operation with them, so that they feel that their efforts are appreciated and supported, and are thereby encouraged to further the good work whenever they can. This is co-operation in its broadest sense.

But we should also initiate it in every community, in every district and valley of this broad State. Let us form local clubs, meet once a month, at least, and exchange ideas. There is a still stronger reason for this than for State Societies, for all the knowledge gained, all the experiments made by us as well as our neighbors, comes directly home to us, and can be applied to our own case, in our own vineyards and cellars. Let us not have any secrets, but act with the conviction that the common good is also the good of the individual ; and have no other ambition but that of applying the knowledge gained, more practically and forcibly than our neighbors. Let the knowledge gained be common property, and the best man win in the common race after perfection.

Still a more intimate system of co-operation can be followed by each neighborhood ; where we can unite by purchasing labor saving implements in common and for common use ; helping one another to prepare our vineyards, by using the same teams in turn to break land, subsoil, etc ; and where there are several small growers, each unable to build cellars and purchase machinery, tanks and casks ; three or four can unite, build a common cellar, of which one of them, perhaps more skilled than the others, can take charge, and work up the product of all. This will lessen the cost to all, they can make a better product, more uniform and in larger quantities, thus have better chances to sell, and cheaper transportation than each could have alone. They will save in purchasing stakes, machinery and casks together; in freight, in labor, in building, and again in selling and transportation. In fact, they can work more profitably in every way. But to do this, there should be perfect harmony, and all jealousy be absorbed in the common interest. The advantage gained by thus combining work, capital, and knowledge, are so apparent and manifold, that they must at once appear to every sensible thinking mind.

Co-operation can also, and ought to exist between the wine maker without means, and the capitalist. Both can combine ; the grower in furnishing good wine, and the capitalist to store it for him, making advances which will enable the grower to work on, and for which the wine offers ample security, until it is aged and finished, ready for the world's market. If capital is thus combined, they can erect ample buildings, engage a competent man as cellar master, offer greater inducements to Eastern and foreign dealers, who can secure large lots of uniform wines, and fully developed goods that will please their customers, and gain a reputation for our products which they could not acquire under the system followed so far.

These are but a few of the many advantages, which a thorough system of co-operation would secure for us. The reader will easily see where it can be made applicable to his case.

CHAPTER XXIII. '

WOMAN IN THE VINEYARD.

I have already taken occasion to allude in the preface, to the warm interest some ladies have taken in our industry, and that some of the best managed vineyards and wine cellars are under the control and personal supervision or women. Miss Austin, at Fresno, planted and managed for years one of the largest and finest raisin vineyards, gained enviable notoriety for the excellence of her products, and although now she has taken a male partner of her joys and sorrows, I do not doubt that her interest and influence is as prominent and beneficial as before. What I now wish to place before my readers is the wide field of pleasant labor for women which our beloved industry opens to them ; a field in which I have no doubt that many will find pleasant change and relief ; while to thousands of industrious women it would offer a more healthy means of gaining an honorable living, than the work in factories, the scanty pittance they can earn with their needle at sewing and embroidering, or the still more unhealthy work of washing and ironing.

Let me not be understood as advising that our fair friends should take the hoe and the plow, or drive the stakes, and do the hard work in the vineyard. These are not for them, and every true man and American citizen will rejoice with me that we live in a country where woman is spared them. But there are many of the lighter and more pleasant operations, which they can do as well and better even than men, as their fingers are more nimble and quick than our more clumsy appendages. Let us consider them in succession; and I think my fair friends will be surprised to see how much they can do in the vineyard; provided always, that they have not more pressing duties at home as wives, mothers, daughters or sisters, but wish to earn their living in a pleasant, profitable and healthy manner.

Let us begin in the winter, when the men are pruning the vineyard and follow it up to the vintage.

First, there are cuttings to make from the trimmings, which the men can bring in for them, and which they can cut easily, with one of the little grape shears described and figured in picking grapes. This they can do at or near the farm house in pleasant weather, and I have seen them cut and bundle from two to three thousand per day. At the usual price paid to men, where they make them by the thousand, 50 cts; they would earn from $1 to $1.50 per day, and I would not pay *them* a cent less than men, because they are *women*, and generally work cheaper; but think them entitled to full pay. Then comes the tying in the vineyard, when it has been pruned, and I am sure that I would be willing to pay them full wages, as their deft fingers will do the work quicker, and generally more accurate than men. Then comes suckering, thinning and pinching the young growth, all easy and pleasant work ; tying of the young growth in grafts, etc., which will continue for several months. If their bright eyes are protected by a pair of goggles, they can also do the sulphuring with a pair of bellows. This will take us nearly to the vintage ; and at grape picking I am certain they can do as much and as good work as men, and should earn as good wages, besides all the grapes they can eat, and a glass of wine occasionally to make their eyes more bright, their cheeks more rosy, and their steps more elastic than they ever will be in the din and impure air of factories, or in the close sewing room.

But many will say : " This may be so, but it is not customary. It is not considered quite stylish, or the proper thing for women, to work outdoors." To these I would say, "I know of hundreds of estimable ladies who work their own flower gardens, water and tend them; and I have never seen the finger of reproach pointed at them for doing so, but they were, if anything, held in higher esteem by their neighbors and friends for thus seeking recreation and pleasure among God's choicest gifts. Throw away this mawkish sentiment, fallacious as it is; and believe that all honest labor, commensurate to your strength, is ennobling instead of degrading ; that "to the pure all things are pure," and certainly there can be nothing wrong in light work, under God's own sky, among the choicest of His handiwork, and in daily and intimate communion with Nature, which will naturally "lead you up to Nature's God." I hope that my fair readers, (and I am vain enough to hope that I will have many) will see in this, not a fanciful and unreal theory, but help, by their example and advice to make it a practical reality. I would not add one single iota of labor to the task of those who already have their full share; I would rather lighten it by giving to each good housewife a pleasant companion and friend, who can help her occasionally, and spend her spare time in the vineyard as indicated ; and furnish pleasant and light employment to thousands, who are now confined in the cities, and inadequately paid for work that will eventually ruin their constitution. I want to bring them to a healthier atmosphere, morally and physically, than they now breathe. And I am confident that my lady friends, with their usual good sense, will aid in the good work.

CHAPTER XXIV.

COST OF ESTABLISHING A VINEYARD, AND ITS PROBABLE RETURNS.

That this must, of necessity, vary very much in different sections of the State, is apparent to anyone at once. We can, however, make a more safe calculation now, when wine and grapes are so low that they can hardly get any lower; than a few years ago, when the prices ruled high, and wine makers were outbidding each other, bringing prices up to an unnatural height, and when prices of wines declined, they must necessarily loose.

We will take for our estimate a raw piece of land, which has to be cleared, and which can be bought for $50 per acre in one of the northern counties :

Clearing land...	$25.00
Preparing land, plowing, harrowing and rolling..............	5.00
600 vines, Riparias or Californicas........................	10.00
Marking and planting......................................	5.00
Cultivation, first year...................................	10.00
" second year..................................	15.00
Grafting, including cost of scions........................	5.00
Staking and tying...	15.00
Cultivation and pruning, third year.......................	20.00
Total...$110.00	

The fourth year, if the soil is of ordinary fertility, the vineyard ought to pay for cultivating and even more. We will make our estimate of returns from the fifth year, when it ought to be in full bearing, and the following years would be about the same average :

```
Five tons of grapes at $12 per ton....................... ........ $60.00
Deduct from this for labor....... .. .................. .........$20.00
Interest on land and capital invested................. .......... 15.00
                                                        -——35.00
                                                        ..- ——
    Net profit.............................................. .........$25.00
```

This, as remarked before, is subject to many variations, according to cost of land, cost of clearing and cultivating, prices realized for grapes, etc. But it may be called a fair average; and the grapes estimated as low as good grapes can be sold and raised. The present prospects are that they will be from $12 to $15 this season in Napa, red grapes bringing the first, white grapes the latter price, and even higher. In other sections, where grapes are still cheaper, they will also produce more to the acre. These are, so I think, bottom prices. If the condensed must industry steps in to make a market for our red wine grapes, and there are more wine storage houses established, it will relieve the present glut, and we will receive better paying prices again. But even as it is, it is better returns on the investment of capital than wheat or grain in general will furnish, and far better than most mercantile ventures will bring.

Raisin making and table grapes pay much better at present than raising wine grapes. Yet a few years ago, raisin makers were down hearted on account of coulure, stagnation in the market, and low prices. Since they improved the quality of their goods, and freight to the East have been reduced, better modes of packing prevail; both raisin men and the shippers of table grapes feel jubilant, and see their prospects brightening every day. So it will be with the wine interest. The present depression is due in a large measure to the inferior quality of the wines sent out. The wine of 1884 was very light, and it was followed by the crop of 1885, which was to a large extent, badly fermented ; but both were bought and rushed East in spring of 1886, during the time of low freight, and the market glutted with inferior wines, which did not pay the shipper, and disgusted the consumer. The crop of 1886 is one of the best in quality ever made in the State. It is a wine that will improve with age, and make a name for our product. The inferior varieties are disappearing, and new vineyards of better varieties taking their place. Our people are learning fast, and we may look for improvement in every respect, quality of product, facilities for shipping, extended markets by new methods; and I think I may safely predict a change for the better.

(Since the above was written, the vintage has passed, and prices ruled higher than anticipated, from $15 to 18 per ton, while the prospects are good for a rapid sale of wine at good prices.)

PART II.

WINE-MAKING IN CALIFORNIA.

CHAPTER I.

It will hardly be expected that I should enter into the practices and secrets of the wine dealer and chemist, in a book which only aims to be the guide of the cultivator, and to render grape culture and wine making easy and practicable for the masses. I shall therefore confine myself mostly to the manufacture of light still wines, and try to demonstrate, in a simple and plain manner, the rules and practices which are necessary for success in this branch. I have made but little sweet wine, nor do I like it; nor do I know much of sparkling wines or brandy. But wine making proper, in relation to still wines, although a very simple art, yet is governed by certain rules which can not be infringed with impunity; any man with sound common sense can become a successful wine maker if he observes them, and also uses that common sense to make due allowance for variations in product, seasons, etc. I shall be as concise and clear as possible; and hope that any one, by following out my directions, will be able to make a good, sound, drinkable and saleable wine, which will be healthy and palatable for him and those who may buy it. My instructions will be calculated more to benefit the smaller cultivators, who work from ten to fifty acres of vineyard, than those who cultivate from 100 to thousands of acres. These can generally afford to build costly wineries, and employ the most intelligent labor, which will certainly be wise economy for them. Still the principles governing wine making remain the same, and I also hope to interest them, as they may find some suggestions which they also can put to practical use.

17

CHAPTER II.

THE CELLAR !

Before *making* the wine, room should be provided to *keep it.* This, in making a small quantity, need not be an elaborate structure; in fact a common house cellar, which can be kept at a temperature not exceeding 80 in summer, will answer, if only a few casks are to be made. The main consideration is always to have the wine *well fermented* and *thoroughly clear* the *first winter*; if this is the case, it will keep almost any where in our temperate climate where we have no severe cold in winter, and the nights are generally cool in summer. I have seen as good wines stored in simple holes in the ground, as have come out of the finest cellars. But if you want to follow it as a permanent business, and make from 10,000 to 30,000 gallons annually, with storage room to keep at least a part of every season's crop, you had better erect good, substantial buildings; whether of stone, brick or wood, is immaterial, and may be governed by the facility you have of obtaining the material. Double walls of redwood are about as durable as stone or brick, at least durable enough to last a lifetime ; the main point is that the lower story at least, in which you aim to store your wine, should keep a fairly even temperature, not varying much from 70°, 65° is still better. At the Talcoa Vineyards, near Napa, which I managed for over five years, we had a building with a capacity of 60,000 gallons, which answered all purposes admirably, and in which I never had a case of imperfect fermentation; I will describe it here, and anyone can change the dimensions to suit his wants.

The building is two stories and a half high. It is built of stone, forty by sixty feet ; the lower story is almost entirely under ground and twelve feet high; not arched, but with a double floor above, which is supported by a double row of strong wooden pillars, twelve feet from the wall. This gives room for four rows of two thousand gallon casks, one on each side, and two in the middle, with sufficient room between for pumping and racking, handling of casks for transportation, etc. It has two rows of six casks on each side, one row of five casks across at the further end, and two rows of five casks each in the middle, making a capacity of twenty-seven thousand gallons, to which can be easily added five thousand more by putting smaller casks on top. It is built into the hill, with a double door, even with the ground below, towards the northeast. The second story has the same dimensions, but is above the ground, with its entrance from the southwest, also double doors, and is used as a fermenting room. It is only ten feet high, however, and contains two rows of casks of one thousand gallons each in the middle, with smaller casks and fermenting tanks on each side; also with a double floor, supported by wooden pillars above. The third story, or rather half story, contains the wine making apparatus, stemmer, crusher and two presses, a few tanks for fermenting white wines in smaller quantities ; boxes and other implements used in wine making, and can also be used as a shop in rainy days, to make cuttings, etc. It also has double doors towards the southwest, and the grapes are handed over a platform from the wagon, from which the approach is even with the floor of the second story. Of course a steam stemmer and crusher could be put up in the upper story, and the grapes run up by an elevator, if desired. All the stories are connected by holes, through which hose can be run from the press or any of the tanks or casks above, into the casks below; and racking from the fermenting vats in third story, to the casks in the second, as also from the second story to the first, can be done without pumping or handling, simply by gravity; also from the presses to the casks below, thus saving a great deal of labor. A staircase connects second and third floors inside. The lower cellar keeps a temperature of about 60° F. summer and winter, the second story varies from 55° to 75°, mean temperature about 70°. As stated before, I have not had a single case of imperfect fermentation there for five years, and therefore consider this, for our latitude and climate, as near perfect as can be, to make good, sound wines.

But, while it should be the ultimate aim and object for every grape grower to make his own wine, let me add a word of caution here to those who, like myself, are not blessed with abundant means. Do not build your cellar and make wine before you are able to pay for it. To build a cellar, and get casks, press and all the necessary utensils, generally costs more than the most minute calculation will show, and it is not safe to run into debt, when you are not fully sure, that you can sell your wine promptly at fair prices and thus pay them off. I have seen too many failures, resulting from apparently safe calculations, to advise you to burden yourself with debts, to secure temporary advantages. Rather sell your crop of grapes to the next winery, as long as you can obtain paying prices ; and then, when you have the money laid by, build a cellar and get casks according to your means, to which you can add as you become able to do so. You can make a few casks for your own use, and store them in a house cellar, or shed even, in our temperate climate ; but do not speculate rashly, and involve yourself beyond your depth, on seemingly safe calculations and inviting probabilities. Your crop may fall below your estimation, or you may not find ready sale for your wine at prices that will pay ; and the result of a good many such ventures has been, that the banks owned the cellars and the vineyards and homesteads also, be-

fore the matter was finally arranged. Go slow, labor patient-
ly and persistently ; and you will not fail to reap your reward;
even if it takes a few years longer ; your mind will not be
burdened by the consciousness of debts, and your work will
thrive and progress better, as long as you are free from embar-
rassment of that kind.

CHAPTER III.

WINE MAKING APPARATUS.

Casks. After the cellar is built, the next thing in order
will be to obtain the necessary casks. We want to be ready
in time for the vintage, so that everything can run smoothly
and without hinderance then. Fortunately we have the
months of July and August here, during which there is not
much to do in the vineyard, and they can mostly be devoted
to work of this kind. Casks of all dimensions are now made
in the State from well seasoned Eastern oak wood, and these
will be the best for permanent use. Large casks save room,
and are the most economical and lasting; though smaller
casks or puncheons will ferment quicker and more thoroughly.
It is evident that a thousand gallon cask is filled and racked
with less time and less labor, than six puncheons of 160 gallons each, while it also takes less room. Therefore it will be
advisable to have large casks for the larger quantities, your
leading varieties ; and have smaller ones for your choicer kind
and smaller quantities. Large casks cost now from 7 to 10
cents per gallon, while common puncheons cost about 6½
cts. The former are provided with doors or manheads, to

facilitate washing and cleaning, so that a man can slip into them, and brush and rinse them from the inside. Oblong casks save room, and are also easier cleaned than round. But we also want frames to lay our casks upon. These can be made from stout scantlings 4–6 inch, the first to be laid 2½ feet from the wall, the second 2½ feet apart from it, and supported by either cross pieces of timber, or better still, pillars of stone every six feet. The inside scantling should be two inches lower than the one next to the wall, so as to give the cask a slight inclination forward. This will facilitate racking and cleaning. If there is a concrete or cemented floor in the cellar, it will also be cleanlier and better.

On these scantlings we lay our casks, giving room to the first one sufficient for a man to pass between the wall and cask. There should also be room enough behind the casks, to allow frequent inspection, lifting in racking, etc. There also should be a space of six inches between the casks, so that each one can be handled and moved independently from the other. For smaller casks, the layers of scantling can of course be nearer to each other. But all should be elevated about eighteen inches from the ground floor, to admit of the free use of utensils in racking. Before the casks are placed, especially if new, they should be made *wine green*, as the general term is, that is the woody taste contained in the oak, should be drawn out. Soaking with cold water, and steaming afterwards, is generally sufficient in smaller ones; but in large, heavy casks, it is well to make some fresh lime water after rinsing with cold water, by adding two pounds of lime to four gallons of water; throwing the lime in first, then pouring in the water, and by rolling and shaking, bring it into contact with every part of the cask. This is about the proportion for a thousand gallon cask. Then wash clean, emptying out all the lime water, and the cask will be ready for use. Sal Soda is also used for the same purpose, and is equally good.

Tanks. For fermenting, we want tanks, made from red-wood, which can also be used for storing wine at an emergency. In fact, many wineries in the State use them altogether, as their cost per gallon is much less, only about 3½ cts. But they are more apt to leak than oak casks, and the wastage will soon amount to about the same, or run beyond it. Besides wine is apt to acquire a peculiar taste from them, which is not desirable. For fermenting, they should be rather wide and low, than high and narrow, as it affords better facilities for stirring, of which I shall treat further on, and the mash does not become so hot. About three and one-half or four feet high, by five feet wide, is a convenient size. They should be of one and one-half inch lumber, smooth and without knot, and also have a top with a manhole, which can be put in when desired. These are placed upright, on a similar frame of scantlings as the casks, in the fermenting room.

Stemmer. This is a necessary implement, and consists of a strong wire screen, (Fig. 26) which can also however be made

FIG. 26.

of wood. Here mostly strong galvanized wire is used, a box

of grapes is poured in, and rubbed back and forth, by a peculiar swinging motion of the two men who work it; the berries
are rubbed off and fall into the frame below, from which they
pass into the crusher; the stems remain, and are emptied out
from time to time.

The Crusher. As generally employed, consists of two
wooden rollers, either plain or grooved, and about eight inches
diameter, so geared with cog wheels that they will run against
each other, drawing the crushed grapes from the hopper above,
as shown in Fig, 27. They are turned by a crank, and set
above a wooden trough, into which the
crushed berries fall, and are emptied
into the press or fermenting vat. The
rollers are so adjusted with screws, that
they can be set so the skin of the berries is broken, but not the seeds; as
they impart a disagreeable taste to the
wine.

Fig. 27.

The Press. This can be made on the old-fashioned lever
plan, and although rather inconvenient in handling, it is
really a good press, as it acts continuously. For this purpose
a hole is morticed into a tree, or if it is strong enough, the
end of the lever may be fastened to the cellar. A long beam,
say twenty feet, is then chosen for the lever, which by a
strong bolt is fastened in the hole, so that it can move up and
down freely. A bottom for the press is then made close to
the end of the lever, where it is fastened to the tree; constructed
of strong two-inch planed lumber, grooved so as to give the
juice a chance to run off, and furnished with a two-inch rim,
to keep the juice from running over. One side is slightly
declined, and a spout of tin of say two inches wide, fastened
to a hole in the rim, where the juice can run out, into a tub.
We now make a box, say 3½ feet square, of strong two-inch

timber, and perforated with half inch holes for the juice to run
out.   This box can be of any desired height, or it can be
made in sections, and enclosed with a frame to give it more
strength.   Three to four feet is about the usual height.   A
strong board, also perforated, to fit into the box as a follower,
completes the arrangement.   When the box is filled with the
crushed grapes, the follower is put on, some cross beams if
necessary; and the lever, which can be worked up and down
at the other end with a rope and pully.   If not found heavy
enough, it can be weighted with stone, and presses all the
time.   This is very simple, and any ordinary mechanic or
" handy man " can make it.

There are of course, many other presses.   One of the best,
of which the smallest size is sufficient to work up from 10 to
20,000 gallons, is the California Wine Press, which can be had
at Woodin & Little, San Francisco.

Smaller Implements.   We want a good many smaller
implements, among which are tubs, to be used for the press
to receive the juice; faucets of brass with threads cut on the
end to attach hose; hose to conduct the must from the press
to the casks, which should be inch and a half diameter outside measure; pails and cans, the last made of strong tin, to
hold about three gallons, narrower at top than bottom, and
with a rim a foot below; and a strong wooden funnell, Fig. 29,

oblong, with a copper pipe in
the bottom at the narrowest
end, and two short wooden
legs, so that it will set firmly
on the cask, with a capacity of
four to five gallons.   The saccharometer and testing tube
have been already referred to,
and are indispensable to work

FIG. 29.

understandingly. The acidimeter, Fig. 30, although useful
at times, is seldom
necessary here, as
our musts, when they
have from 22 to 26
Balling, have no ex-
cess of acids, and
they should not be
lighter to make good
wine. Besides, it is
more difficult to use
the acidimeter prop-
erly, as it can only
show the acids cor-
rectly when fermen-
tation has drawn
them from the skins
and flesh, therefore
in inexperienced
hands generally does
more harm than
good. We also want
a strong pestle or board, say nine inches in diameter, with a
five foot handle attached, to stir the pomace in the fermenting vat.

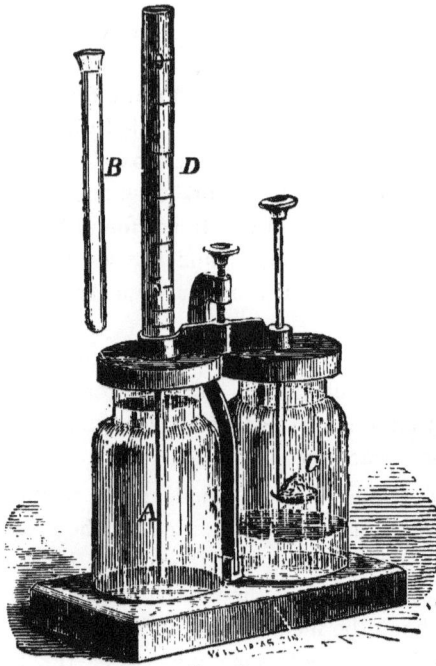

FIG. 30.

This, of course, refers only to small establishments, who
want to do the work by hand. Those who intend to work
on a larger scale, by steam or water power, will do best to get
the combined stemmer and crusher, as well as other machinery, from L. Heald, Crockett, Contra Costa Co., who makes
a specialty of wine making machinery, presses, etc., and
whose machinery is used by nearly all the large wineries in
the State. If they will state their wants and the extent of
their operations, he will furnish just what they want.

CHAPTER IV.

MAKING DRY WINES.

As I have already given directions for picking the grapes, and the proper time and manner when to do so, we are now ready to make our wine, as soon as our apparatus is all clean and in working order. Dry wines may be divided into three classes, all requiring different treatment, and I shall consider them in succession.

(a) WHITE WINES PROPER.

This comprises all wines made from grapes which contain none or very little coloring matter, in short, all wines resembling Hocks or Sauternes, the two best known types of German and French white wines. Let us define these two classes a little closer.

Hocks we call the wines from the Rhine, the Moselle, the Palatinate, and other German and Austrian provinces, which are characterized by their light yellow or greenish color, sprightliness and agreeable acidity, as well as their agreeable bouquet. The majority of them are light rather than heavy, and it is considered a fair proportion of alcohol if they contain eight per cent. They are preeminently the wines to "make glad the heart of man," the main ingredients of the famed "Maitrank," which have furnished the inspiration for the innumerable songs in their praise for which the "Vaterland" is famous, and which have given the Rhine its fame as the most poetic and romantic stream on earth.

Sauternes are generally fuller, softer, and smoother than Hocks, and many of them are slightly sweet. The far-famed Chateau Yquem is the representative of that class, the noblest white wine which France produces. But while France makes some very choice white wines, her fame was mostly gained by her red wines and clarets. We may therefore safely assert that the French are preeminently a claret producing and consuming nation, and the Germans the consumers of white wines.

Of the Hock type we have the Riesling family, including Chauche Gris, Green Hungarian, Traminer, White Elben, Yellow Mosler, and for a lighter type of Moselle wines, the Gutedel and Burger.

Of the Sauterne type, including the hermitage wines, Marsanne, Roussanne, Sauvignon Blanc, Semillion, Muscadelle de Bordelais, Pedro Ximenes, (generally called Sauvignon Vert).

The Clairette Blanche, although French, and the Herbemont, American, I should class with the Hock varieties, on account of their sprightliness, pleasant acidity and bouquet.

For wines developing a sherry type naturally, Palomino and Sultana. It will naturally be inferred that only such varieties should be blended or mixed together as belong to each type.

When our grapes are stemmed and crushed, they should either be pressed at once, if the light, greenish color now in vogue is desired, and a very smooth delicate wine; or if a wine of golden yellow color is the preference, they may be put into the fermenting vat, and allowed to remain over night, say twelve hours, on the skins. This will give the wine a deeper color and more flavor and character, though not so smooth and agreeable; I know that I come into conflict here with Mr. Arpad Haraszthy, who is generally considered authority, and discourses on fermentation at every meeting. He advises to leave them in the fermenting vat for three or four days; in short, treat white wines about as red wines should be treated. I do not pretend to be an authority, but I also do not acknowledge any; and my practice has taught me that white wines, thus treated, loose all that delicacy, smoothness and sprightliness which to me constitute their chief merit, and become rough and acid. But my readers can easily satisfy themselves, by trying both ; than follow the method which suits their taste best, and gives them the most saleable wine.

The must can be run into the casks from the press at once. Of course these should be "wine green" beforehand ; if new, they should be made so as indicated before; if old, and they have contained wine before, they should be made perfectly clean and sweet by thorough washing. The rinsing chain, (Fig. 31) is a good implement for loosening any impurities,

FIG. 31.

and the smell will indicate whether they are clean and sweet. If at all mouldy they should be rinsed with lime water or sal soda, and if they lay empty for several days, they should be sulphured as soon as dry, and the bung drove in tight. It will be found very useful to burn some good pure grape brandy in them, which will fumigate them thoroughly. Take about a quart to a thousand gallon cask, pour it in at the bung, and ignite by dropping a burning match into the brandy from above. As soon as the brandy is burning well, lay a cloth over the bung; but do not drive in the bung until it is all burned, as it may burst or injure the cask. This will penetrate the wood deep enough to benefit the must, and is much better than heavy sulphuring. If you do sulphur, rinse the cask before filling, to take out the smell and taste, which the must easily acquires, and which is also deleterious to active fermentation. Fresh brandy or whiskey casks may be used for fermenting wines, provided there are no other flavorings used in the liquor, such as gin, chervil, anisette, etc. But they will not do for fermented wines, or for racking. The quantity of brandy to be used will of course correspond to the size of the cask. Fill your casks with the must to within about eight inches of the top, to prevent its running over, and divide the pressing which runs last, equally

among them, if you have several to fill, if you want to make
a uniform wine. Some prefer to have the first run by itself,
and fill the pressed wine into a separate cask. The first will
of course make the most delicate wine, while the last will be
more harsh and rough, from the tannin and acid extracted
from the skins and seeds. But this tannin is generally needed
to clear the wine and make it durable, and if pressed as soon
as indicated, there will not be an excess of it. The bunghole
may then be covered by a fresh grape leaf, to keep insects or
dust from entering, and the bung, or a small sack with clean
sand laid on, to keep it there until the wine has become quiet.
To fill up, some must of the same kind should be filled into a
smaller cask ; and when violent fermentation is over, say in
five or six days, they can be filled up to the bunghole. As
soon as fermentation is over, which you can tell by holding
your ear above the bunghole, by the absence of the hissing
and seething noise which accompanies fermentation, the bung
can be put in, at first lightly, and after a few days, it can be
drove in tight.

A great improvement on the solid bungs for the management of young wines is the perforated bung. For this, good
spruce, maple or ash bungs are used, made about four inches
long, tapering gradually. A hole with a half inch or three-
eighths augur is then bored through them the whole length,
and filled with cotton steeped in salicylic acid, pressed together solidly. This gives enough vent to the young wine,
yet will act as a filter to the air when the wine becomes quiet
and fermentation ceases. They can also be used for casks
and barrels that are kept "on tap" as the phrase is, for a short
time; and though I do not advise their use for an unlimited
length of time, yet they are a better preservative than solid
bungs, which can hardly be closed enough to prevent all access
of air. In this case the air is freed from all impurities, and
will not vitiate the wine. To young wines, it prevents the ac-

cess of impure air, yet gives them sufficient vent to finish fermentation completely.

The treatment of all these varieties is about the same in the first stages of their development, if they are to be dry, light wines. The German Muscateller or Frontignan can also be made into a dry wine in the same way, but it is best to press it immediately in all cases, as the flavor of the grape is naturally strong, and will be developed to an unpleasant degree by fermentation on the skins.

(b.) MAKING WHITE WINES FROM BLACK GRAPES.

This is often advisable, and especially now, when white wines sell so much more rapidly, and at higher prices than reds. We have some red varieties however, the so-called "coloring grapes," which have red juice, and which therefore cannot be used for white wine. These are the Lenoir, Teinturier, Gamay Teinturier, Pied de Perdrix, Petit Bouschet, and perhaps a few others. In fact, anything of which the juice runs red, when you squeeze a berry between your fingers, cannot be used for *white wine.* But when a vintner has Mission and Malvasia yet, he certainly ought to make as much white wine from them as he can, and either throw away the pomace or use it for distillation or port. They make a fair white wine, but the dry red wine made from them, deteriorates with age, and never is very good. But there are many red wine grapes, which have all the color in the skins, and which, when pressed immediately and lightly, will make a nice white wine from the first run, when the mark or pomace, which of course contains a good deal of juice yet, and all the coloring matter and tannin, can be put into the fermenting vat, either with other red grapes, or fermented alone. Of course, the first run is the best juice, and will make the smoothest wine ; but this is not generally sought for in clarets; in fact, the trade has so far demanded deep color, astringency and flavor, all of which would be increased by taking off the first run immediately after crushing ; and fermenting the remainder for five or six days.

Among the grapes which make *choice* white wines if pressed lightly are Chauche Noir, Meunier and Carignane ; the Zinfandel, Blaue Elben, Mataro, Beclan, Grosse Blaue, Mondeuse and Petit Syrah can all be utilized in this way, and make very nice wine. The Herbemont should also be treated in this way, as well as the Rulander and they make delightful white wine. Press very light, and quit pressing as soon as the juice assumes a red tinge, using the remainder for red wine. Treat the " first run " just as the white wine proper, and you can thus increase the quantity of white wine, if it should be desirable. If you have fresh pomace at same time, of Marsanne, Riesling or any of the choice Sauterne or Hock varieties, and will ferment your must from red grapes on them for a single night, you will find that you can give it the character of that special variety, and thus utilize your product much better. I have thus fermented white Malvasia on Marsanne and Pedro Ximenes pomace, which would pass for very fair Hermitage, although of course, it could not be called a " grand vin." There is a wide field of useful experiment open for us in this direction, and it certainly is legitimate blending, the highest art in viticulture. It seems to me there is a more paying field in that direction than to try to improve and ameliorate our clarets by cutting them with light white wines, to make them acceptable to the public taste.

(c) RED WINES.

In making red wine, we have of course a different object in view. In white wines, we desire sprightliness, delicacy, smoothness and bouquet; in red wines, we want good color, and astringency mainly, and in this State, even the fine bouquet, which ought to characterize good red, as well as white wine, has often been sacrificed to attain the two first, and the mellowness which a good red wine ought also to have, is seldom found. While I fully recognize the importance of good color and astringency in red wines, I hope to show my readers how they can make them with a sufficiency of these, and also have them of good bouquet and mellow. The general practice in this State has been formerly, to crush the grapes, put them into six feet high fermenting tanks, and then let them work out their own salvation. I have often been in wineries that looked more like slaughter houses, with the purple juice bubbling over the top, a crust of a foot thick had formed on top, which had become dry and mouldy, was swarming with vinegar flies, and in many cases, maggots were crawling around lively. When the fermentation was over, the whole mass was often left for a week yet, as the manipulator thought to gain for it color and tannin, and become more saleable thereby. That under such treatment decomposition and acidification had often set in, can hardly surprise any one. Yet these were often, and even in the majority of cases, so called "old skillful wine makers" from France or Italy, who felt insulted if their practice, which their fathers and grandfathers had followed before them, was not considered perfect. If told that five days of thorough fermentation would extract all the color and tannin, and make a deeper colored and more lively wine than theirs, they would not believe it, had it been demonstrated before their eves. I know of large establishments, even now, which make from 200,000 to half a million of gallons of Claret every year, who keep their fermenting rooms at such a temperature at night, that suffocated rats are strewn about the floor in the morning. And yet the wine made under such conditions is sent all over the Country as " California Claret." Is it a wonder that it has a bad name and reputation?

Most of the clarets are now fermented under so called "false" or perforated tops; that is, after the crushed grapes are put in, a perforated top which fits on a rim or cleats nailed to

18

272 GRAPE CULTURE AND

the inside of the tank, is put over them, so that the juice comes through the holes and stands above the mash and the top fastened there. This is better than the first method, but yet admits of great improvement. Naturally, fermentation commences above, where the fluid comes into contact with the air, slowly progresses downward, so that when the top has already passed through fermentation, and become bitter, it is still sweet below; and unless the lower fluid is drawn off and poured on above, there is a great difference between the top and bottom in its development, and it is very difficult to tell when it is ready for pressing.

My method, by which I have always made good clarets, and had them ready for the press within five or six days, is as follows :

After the grapes are crushed, I fill them into the fermenting vat. I prefer rather shallow and wide to the deep and narrow ones which have been in use generally, say four feet high by five feet wide. A small screen of wire should be fastened over the faucet hole to keep out the skins and stems if any are left, and to let the liquid run off clear. Then fill your tanks to about a foot from the top, so that they will not run over in fermenting, and stir the mass at least three times a day while fermenting, with a wooden pestle made for the purpose, taking care that the whole mass is well mixed, and the skins rising to the surface are submerged again, so that no acetification can set in. This is a process of aerating, by which the skins which have been exposed to the air are again submerged and keep up a lively fermentation ; but if the temperature of the fermenting room does not rise above 75 or 80°, as it ought not to do, the temperature of the mash will not rise above 90ᵘ. It will be well, should fermentation get violent above, to test the liquid below, at the bottom of the vat, and if that should be much below the temperature at top, to draw out some by the faucet below,

and pour it on top.   This will equalize and aerate the whole,
and within five days we generally find that all the sweetness
has disappeared.   The test should be taken from the bottom
of the vat, and if that does not show any sweetness we can
depend upon the whole being ready for the press.   It is a
mistaken idea to think that the must will gain any color or
tannin after it has become cool and quiet, for the color will
remain fixed better after the cooking process (and fermentation is nothing else) has subsided.   If the wine (and it has
gone into the vinous state as soon as fermentation has done
its work) remains any longer on the skins, it simply loses
all the freshness, and fruity flavor which are so necessary; in
fact constitute its chief value to the buyer.

To comprehend this, we only need to look a little closer
at the nature of fermentation, and what it will accomplish.
During the process, carbonic acid gas escapes all the time,
rising in bubbles to the surface, and if the mass is stirred and
kept submerged, no acetification can take place.   But as
soon as it has become quiet this ceases; the surface is still,
no carbonic gas escapes, and the exposure to the air is bound
to affect the young wine to its positive injury.   To test this,
only leave it quiet for one or two days, and you will see
mould appear on its surface, that sure sign of putrefaction
and decomposition.   It robs the young wine of all that liveli-
ness and fine aromatic properties it ought to possess, and
which we find developed in our white wines to a much
higher degree than in the red, just because it is not exposed
to this putrefaction process.   Yet we find old wine makers
advocating the practice of leaving the wine on the skins long
after fermentation has ceased, to gain color and tannin, *as
they suppose*; while it accomplishes just the contrary; for the
exposure to the air after fermentation has the tendency to decrease the color.   And even if this were not the case, I
would rather have a wine with a trifle less color, and more

life and bouquet, than vice versa. The best and choicest French and European clarets are not noted for their depth of color especially, but for their fine bouquet. This gives them their value, and it is this in which the Italians, with their naturally excellent product, are so far behind the French. They are, perhaps, the most slovenly wine makers on the face of the earth, and today prefer the Mission and Malvasia, with their deep color and roughness, to the choicest red wine grapes. Anything that will " scratch the throat," which is rough and acid is good for them, and their wineries, where everybody makes his own wine, are enough to shock the sensibilities of any common man or woman, and create a prejudice against the use of wine. Let me not be understood, however, as if there were not good and skillful wine makers among them. But they are like angels visits, " few and far between," and deserve all the more credit as honorable exceptions to the rule.

Italy, as recently illustrated by Dr. Springmuchl, possesses some of the choicest material in the world, yet it is generally spoilt for want of proper management, and needs the manipulation of the French, *the* great wine doctors *par excellence*, to make it drinkable, and to introduce it to the world as choice Bordeaux claret.

The method above given is what the French call making claret with *foulage* or frequent stirring, and, if closely watched, will I think produce our choicest clarets and Burgundies, as it will preserve all their freshness, and yet extract all the color and tannin as well as the aroma. That a claret or Burgundy can never be as delicate as a fine white wine will be evident from the above. I have generally made some Zinfandel for my own drinking, which I prefer to the darker colored and rougher wines, and which has all that fine raspberry flavor so characteristic to the grape. To make this I simply draw some must from the bottom of the vat, after it has fermented about thirty-six hours, and fill it into a separate cask. This is much smoother, of a lighter color, being what we call a dark Schiller wine, and to my notion far surpasses all the deep colored wines which I get from the same tank after fermenting three days longer. But it would be no wine for the merchant, who generally buys pale clarets enough, and needs deeper colored wines to give *them* color and astringency. Besides, the taste which requires red wines seeks a different product. It wants color and astringency, and we may as well gratify it, but we ought not to do so at the sacrifice of all the finesse and flavor of which red wines are capable. If we attain a happy medium, giving them deep color, astringency, and a pleasant acidity, they will surely not object to a good share of fine bouquet.

The record of experiments by Prof. Hilgard at Berkeley, given below, of nine different modes of fermentation, will be of interest to the reader, showing what the difference is in the various processes. The only criticism which I wish to make is that the pressing was delayed too long. I think it was a positive injury to No. 559 to delay the pressing until after the sixth day, while No. 560 should have been pressed on the fifth day, and none of them ought to have been deferred after the ninth day. This is the mistake, in my opinion, alluded to before. When active fermentation has ceased, the wine has become dry; every hour delayed works a positive injury to the wine, against which we ought to guard. This is *my* comment on the wholesale deduction against open fermentation made by the learned professor.

My readers will perceive that mine is not the "lazy man's" process. In fact, I do not count on laziness. Those who intend to succeed in this industry must be content to work sometimes all night; if necessary; they must be willing to watch their vines and their wine with unceasing diligence and care, as they would their children; and "Excelsior" must be their motto, even if attained at some physical sacrifice.

Following the very interesting experiments of Professor Hilgard, I give a treatise by R. D'Heureuse on air treatment, communicated to the *Grape Culturist*, (a paper I was then publishing in the interest of wine making and grape culture in 1870). It will serve to show that the same principle, aeration, underlies all rational fermentation. It contains much food for thought, and a great deal of sound theory, which has since been verified in other branches of industry. That air treatment would be very beneficial to our California wines, which contain a large amount of gluten, is evident; and all modern writers on the subject advise it in some form or other, either in the racking of wine, by letting it run through a faucet in a circular manner, through a rose, to bring each particle of the young wine into contact with the air, or by infusing air into the must by pumping through perforated hose. Many predictions of poor D'Huereuse, who shared the fate of most inventors during his lifetime, have since been verified, and it affords me pleasure to contribute thus towards "keeping his memory green." Small were his thanks and his emoluments for bringing more light to bear upon this important question, but the future may yet vindicate his pet theory. It seems like prophesy to read his predictions about concentrating must, shipping of grapes, centralization of wine industry, etc. The same great problems which we are trying to solve, are ably forshadowed in his treatise, and seem to reach fulfilment now.

EXPERIMENTS ON METHODS OF FERMENTATION.

In view of the great interest attaching to the determination of the effect of various methods of fermentation upon the resulting wines, a series of experimental fermentations with one and the same kind of grape, treated differently both in respect to temperature and the appliances used, was carried out with the results given below, so far as the record reaches at the time of going to press. While in some cases the differences in the outcome are already apparent, and are even indicated by the chemical analyses, in others they are at present very slight, and if existing at all, will become obvious only in the development of the wines.

As will be noted, these experiments were all conducted within the limits of temperature adapted to "high fermentation," since no means were at hand for maintaining a temperature sufficiently low for the "low fermentation" proper. It is intended to arrange for such experimental low fermentations next season, in order to test the question whether in wine-making the same precautions now universally practiced in the case of beer, might not be profitably employed in the preparation of wines which, like those of the Rhine and Moselle, are essentially low-fermented and can not be successfully reproduced by the aid of high fermentation. For the present. the question of how best to manage the fermentation at the prevailing vintage temperature of California is the one having the greatest practical interest, and to this end the experiments were directed.

On account of the heavy pressure of vintage work, these somewhat laborious experiments had to be deferred until after the main crop was in, and out of hand. The grapes used were a very good article of second-crop Zinfandel, courteously donated for the purpose by Mr. J. Gallegos, of Mission San José. About one and one-sixth ton of these grapes was received in excellent condition, having been packed in the basket crates sent from the University for the purpose. The berries were rather small and the bunches quite loose, but thoroughly sound; taste agreeably sweet, and juice abundant. The composition of the latter was as follows:

Solid contents by spindle...21.05
Sugar by copper test...19.75
Acid .... .65
Ash .... .27

Nine different samples were fermented, under the following conditions:

(A.) In a hot chamber, kept at a temperature ranging from 95° to 102°, two batches of about 63 pounds each, one (No. 557) left entirely open in the tub, the other (No. 556) covered with a "floating top" that rose and fell with the pomace, screening it from access of air. Both were stirred.

(B.) In a room kept at a temperature ranging from 72° to 75°, five fifty-gallon tanks, each charged with about 230 pounds of grapes, filling the tanks to within ten inches of the top, and arranged as follows:

No. 558. Mash put in in three successive portions, and each kept to itself by means of a lattice framework kept in place by wedges, thus forcibly keeping the pomace submerged and divided into three separate portions, according to the system of Perret; the uppermost frame being about two inches below the surface of the must before fermentation; a cover laid on top of the tank, according to Perret's precept.

No. 559. Mash put in at once and the pomace kept submerged about two inches below the surface by means of a single "Perret's" frame, according to the practice prevailing, to some extent, at Fresno and elsewhere; no cover of any kind.

No. 560. Mash left uncovered and subjected to frequent "*foulage*," or stirring, at least three times a day during fermentation ; a common French practice.

No. 561. Mash covered with a "floating cover," rising and falling with the pomace, and stirred three times a day, washing off the upper side of the cover in case of frothing over ; a method also used in France and adopted as both good and convenient, in the Viticultural Laboratory.

No. 563. Grapes put in whole, stems and all, to be gradually crushed by means of the cross-peg stirrer ; a method much in use in Burgundy, and also practiced at several wineries in this State ; no cover ; stirrer used energetically three times a day. " Morel process."

No. 564.   A tub charged with about 140 pounds of mash and then left to itself, cap, vinegar, flies, and all, without stirring or cover ; the old Californian method.

(C.)   In the cellar of the laboratory, kept at a steady temperature of 62°.

No. 562.   Fifty-gallon tank charged like the rest, with 230 pounds of grape mash, provided with a " floating cover," and stirred three times daily.

In all these vessels the temperature was read off three times daily, during the height of fermentation every two or three hours, and in the tanks provided with the frames the temperature of the top liquid, and of the pomace beneath each frame, was separately ascertained, in order to follow the exact course of the fermentation.   Observations similar to the last were made every morning in the tanks subjected to stirring, so as to ascertain the temperature of the top and bottom layers of the pomace cap formed during the night, and that of the liquid beneath.

The observations made with the several fermentations are plotted in the table below, for greater facility of obtaining a comprehensive view of the results.   Where several figures are placed opposite one and the same hour of observation, they are to be understood as representing the temperature of the top and bottom, if two ; if three, top, middle, and bottom, respectively.   During the first and last stages of the fermentation, when the changes were very slow, observations are sometimes omitted.

The highest temperature observed in each case is printed in full-face type.

It is seen from the table that the high-temperature fermentation, No. 556, went through with extraordinary rapidity, the young wine being dry within two days of the setting of the mash. The same quantity of mash, set without cover and at the outset left without stirring (No, 557), was markedly slower in its course, although the maximum temperature reached was the same and occurred about the same time as in the other case. As the fermentation seemed nearly ended on the evening of the third day, a floating cover was put on in order to prevent acetification during the night, and the hot chamber was opened so as to share the temperature of the room, viz., 75°. But in the morning a cap had formed and a slight fermentation was still going on, as is evidenced by the temperature having remained at 93°, despite exposure to a much lower one. But a few hours later all appearance of fermentation vanished. It is not easy to see why the absence of the cover should have made so much difference in the time of ending the fermentation. Less frequent stirring was probably the main cause.

The relatively small mass concerned in these fermentations prevented the temperature from rising so high as to injure the yeast, 102° being the maximum observed. With the larger masses used in the other experiments, the temperature rose as much as 20° above that of the room ; and correspondingly the maximum in *these* two fermentations would probably have been about 120°, had the same amount of grapes been used.

In the two fermentations (Nos. 558 and 559) with *frames* to keep the pomace submerged, the record shows that while up to the time of the maximum, the temperature was always highest at the top, shortly afterward this relation became inverted, the lower portion being found warmer than the upper. This fact is most apparent in the case where the single frame was used (No. 559), in which the maximum temperature of the must below the pomace cap was actually attained about thirty hours later than in the pomace itself, showing that the fermentation in this lower portion was far behind that in the upper. This consecutive occurrence of

maxima explains why the highest temperature found in the
single-frame process was considerably (4°) behind that
observed in the three-frame process (No. 558), where the
whole mass reached its most intense action simultaneously,
although a slight occurrence of the reversal of temperatures is
observable here also. It should be noted that (as stated in
the table) the latter process was not entirely normal in its
course, a portion of the pomace that should have stayed
below the frames having been carried through the meshes to
the surface by the ascending gas, thus equalizing the temperature throughout the mass much more than would have been
the case had the pomace cap been of sufficient thickness under
each frame, or the meshes of the frame fine enough to prevent the skins from rising to the surface.

Where the single frame only was used, scarcely a grape
skin was seen on the surface ; and except toward the last, the
maximum temperature was always found near the lower layer
of the pomace cap. During the last two days a white scum
was seen to gradually form on the surface of the wine, and in
the end gathered into white mould islets, as was observable
under the microscope. No such scum was to be seen on the
three-frame tank, which had remained covered during the
whole process, except while the temperature was being taken.

Marked differences in the course of their fermentation is
also apparent in two tanks that were subjected to *foulage*, or
stirring, viz. : Nos. 560 and 561. The one which was left
open to the air, and also received an extra amount of stirring,
fermented with a violence greater than any of the others ; so
that in order to prevent a wholesale running over of the froth
and serious loss, it became necessary to fasten down on it a
cover for four hours. The maximum temperature of 95° was
not, however, reached until about eighteen hours later, although the pomace cap at the depth of seven inches showed
101° after the violence subsided. The action and tempera-

ture then rapidly declined, and the mash was ready for pressing quite twenty-four hours before the tank which had heen
fermented with the cover on.   The latter reached its highest
temperature about the same time as No. 560, but it was only
92°, 3° lower, and the most violent fermentation occurred
about eight hours after the violent outburst of the companion
tank was over.   A slight action continued quite twenty-four
hours longer in the tank fermented with the cover on.

In the latter respect we have here the reverse of what occurred in the hot chamber, where the mash having the cover
on went through most rapidly.   But this was not much stirred
at first, and the larger experiment conforms to the presumption in the premises, which is that the more perfect aeration
will bring about the most vigorous fermentation.

In the case of No. 563, the "Morel process," in which the
grapes were put in the tank with the stems uncrushed, and
were gradually crushed with pole provided with cross pegs, the
course of the fermentation seems to have been governed more
by the fact that the stems kept the pomace diffused through
the whole mass, than by the intended gradual crushing of the
grapes.   The latter were so tender that after the fourth day
little more crushing could be affected, the whole having become so liquefied that the berries remaining uncrushed evaded
the pole; but instead of forming a solid cap on top, the pomace
and stems always reached within six to nine inches of the bottom of the tank; and thus the fermentation was accomplished
nearly under the same conditions as that of No. 558 (three
frames), but with the addition of aeration.   Hence the temperature rose higher than in any other mash fermented in the
same room, viz.: to 97°.   This maximum was reached about
the same time as in the others—on the fifth day; but the fermentation continued slowly, and doubtless in consequence of
the occasional crushing of fresh berries, the wine was longer
in getting dry than any of the rest.   It thus appears that in

the actual practice of this method, the effect on the temperature will depend greatly upon the nature of the grapes so treated. The small-berried, thick-skinned Pinots, to which this
treatment is chiefly applied in France, will in general gain the
benefit of a slower fermentation, but in application to such
grapes as Zinfandel, Charbono, and similar delicate-skinned
grapes, the practice seems to present no advantages. Unless
a strict measure is observed in the pounding, in the case of
very juicy grapes, a certain proportion is sure to escape crushing altogether.

No. 564. The "go-as-you-please" method of many early
and some contemporary wine-makers, in which the pomace
was allowed to rise to the top and stay there to the end, exposed to air, mould and vinegar flies, was, of course, intended
only to illustrate "how not to do it." The pomace-cap was
for most of the time emerged from one and a half to two
inches above the must, and began to acetify so soon as the
violent fermentation was over; the temperature in the pomace
rising as high as 89° on the fourth day. But in the absence
of any stirring-in of the pomace, the fermentation in the must
below was slow in completing itself, and a slight action continued into the eleventh day. By that time a generation of
vinegar-fly maggots had developed and was making the
emerged pomace look very lively; in pressing, some animal
juices inevitably mingled with the wine, but the latter showed
no obvious defect at the time of pressing, and its taste was
that of a more advanced product than any of the others. Its
subsequent history remains to be seen.

No. 562 was fermented under precisely the same treatment
as No. 561 (that is, with floating cover, and thrice daily *foulage*), except that the temperature was, on an average, 13°
lower, that is, 62°. It started slowly; its maximum temperature was reached about twelve hours later than in those fer-

mented at 75°, and did not exceed 83°. It had become fairly still on the tenth day, and should have been pressed on the eleventh; while the tank similarly treated at 75° was pressed on the eighth day. It was a healthy steady, fermentation, at no time threatening to froth over the tank, and only for a short time frothing over even the floating cover. It was by far the most comfortable fermentation of the nine.

COMPOSITION OF THE WINES.

The table below shows the composition of the wines resulting from the several fermentations. They were all analyzed, and their color determined, within a few days after pressing, the murk being filtered for the purpose.

COMPOSITION OF GRAPES AND MUST, AND WINES PRODUCED THEREFROM BY DIFFERENT METHODS OF FERMENTATION.

Number		GRAPES.					
		Weight of Grapes in Pounds.	Percentage of Pomace.	Percentage of Stems.	Gallons of Murk.	Gallons, per ton of Grapes.	
556	Foulage, with floating cover .... } T. 96°—100°	61.0	11.1	7.0	5¼	172
557	Foulage, without cover.... .... }	65.0	10.8	7.0	5½	169
558	Three Perret's Frames....................	230	....	......	......	......
559	Single Perret Frame......................	230	....	......	......	......
560	Foulage, no cover...................... } T. 72°—75°	230	....	......	......	......
561	Foulage, with floating top................	230	....	......	......	......
563	Morel Process.............................	230	....	......	......	......
564	Old Style, no cover nor foulage.......	140	....	......	......	......
562	Foulage and floating top, Temp. 62°.......	230	....	......	......	......

	MUST.							WINE.					
Number	Solid Contents by Spindle	Ash	Grape Sugar	Fruit Sugar	Total Sugar by (Copper Test)	Acid as Tartaric	Alcohol. By Weight	By Volume	Acid as Tartaric	Tannin	Color. Intensity	Tint, Purple-red.
556							7.78	9.73	.49	.14	45.4	...2d
557							7.78	9.73	.59	.13	42.4	...2d
558							7.78	9.73	.49	.10	30.0	...2d
559							7.23	9.00	.51	.10	27.8	...3d
560	21.1	.27	16.40	9.35	19.75	.65	7.78	9.73	.65	.10	46.4	...2d
561							7.78	9.73	.53	.12	47.0	...2d
563							7.16	8.93	.67	.13	47.6	...2d
564							7.09	8.85	.56	.12	46.6	...2d
562							7.78	9.73	.56	.12	34.1	...3d

ALCOHOLIC STRENGTH.

As regards, first, the alcoholic contents of the several wines, it will be noted that the same percentage was obtained in six out of the nine ; while three, viz., Nos. 559, 563, and 564, corresponding respectively to the single frame, Morel, and "old-style" processes, show a deficiency which does not differ widely for the three, being not quite one per cent.

In two of the above cases this result was to be expected, and the causes are not far to seek. In the single-frame process, a relatively thin layer of liquid was exposed to the air, constantly agitated by the gas coming from below, and heated by its position just over the hot cap. The alcohol simply evaporated from this isolated portion of the wine, and where this mode of fermentation is practiced on the large scale, I have sometimes found this layer so warm that toward the end of the fermentation the bulk of its alcohol was gone and it had a vapid, flat taste, often more of vinegar than of alcohol.

In the case of the old-style process, also, it is easy to see where the loss of alcohol occurs. It is here the hot pomace cap, offering a large surface to the air and kept drenched with the fermenting liquid by the bubbling up from below, which assists the evaporation. That the latter is accompanied by its transformation into vinegar is apparent to the nostrils so soon as the first violent stage of the fermentation is past.

In the case of the "Morel process" the cause of the loss of alcohol is not so obvious. It might be accounted for by the abundant stirring and high temperature, and, doubtless, this contributes to the evaporation, so much the more as the tems, more or less emerged above the surface, afford better opportunity than a cap formed of skins alone. Yet the loss appears to be greater than can be accounted for on this basis alone, for the reason that in No. 560, where the *foulage* was nearly as diligent as in the "Morel" tank, and which was also open to the air, the alcohol percentage is not sensibly diminished. It is possible that from some cause a part of the sugar may have been converted into some other compound than alcohol; among these, glycerine suggests itself, but the determination of this substance in the wines has not yet been made.

A somewhat unexpected result is the fact that the two hot fermentations (556 and 557) yielded the same amount of alcohol as those fermented at a much lower temperature. The obvious explanation is, that the short duration of these fermentations balanced the influence of the high temperature as compared with those in the slower fermentations, in which the opportunity for evaporation lasted longer. It will be highly interesting to compare, hereafter, the other products formed alongside of the alcohol in the three sets of fermentations.

ACID.

As regards, next, the acid of the several wines, it is not unexpected to find that the open *foulage*, No. 560, on the one hand, and the Morel process on the other, having given the highest figure, the one because of the constant access of air, the other from the same cause, in addition to the extraction of acid from the stems.

The lowest figure for acid (.49) is given by Nos. 556 and 558, the hot fermentation with cover, and by the one with the three submerged frames. In the case of the latter this was to be looked for, and is precisely one of the chief advantages claimed for Perret's method. In the case of the former it is somewhat unexpected, and is the more instructive in contrast to No. 557, the hot fermentation in which no cover was used, and in which the acid is one pro-mille higher. Almost precisely the same difference occurs in the fermentations made at the lower temperature, one with the floating cover on (No. 561) and the other (No. 560) without cover. The beneficial influence of the cover in preventing the formation of acid during fermentation is therefore placed beyond question.

It should, however, be added, that in none of the fermentations made, there is at this time (November 24), a notable amount of volatile (acetic) acid. This is true even of No. 564, the "old-style" one, in which the odor of vinegar was abundantly obvious before pressing. It shows the odor of vinegar plainly in boiling, but the amount is at present less than five thousandths of one per cent.

It is somewhat remarkable that the fermentation of 562, made at the lowest temperature, should yield a relatively high proportion of acid, exceeding that found in the fermentation made under the same conditions at a higher temperature. Whether this is to be accounted for by the longer duration of the low-temperature fermentation, remains to be investigated.

TANNIN.

Considering, next, the matter of tannin, we note at a glance the influence of the high temperature in aiding a complete extraction. The two hot fermentations, Nos. 556 and 557, have given the maximum of tannin, despite their short duration; more even than in the case of the tank with diligent open foulage, and as much as the Morel process, stems

19

and all, which was continued for eleven days; the effect in this case is so marked as to leave no doubt of the influence of this factor, and in it lies, probably, at least a part of the explanation of the fact that the hot parts of our State have yielded more tannin in their red wines than the cooler ones.

The two tanks in which the frames were used (Nos. 558 and 559) present a curious problem. In both cases the same amount of tannin was taken up, although in the one, the pomace was in a solid mass, and in the other, was kept diffused all through. The result is disappointing as concerns the three-frame process, and shows clearly why, despite its apparent advantages, this method of treatment has not been widely adopted even in France. It is evident that simply keeping the pomace in the liquid cannot replace the grinding and disintegrating action of the direct stirring or *foulage*, so far as the extraction of tannin and color are concerned; for a glance at the color-column shows, that the deficiency of tannin is accompanied by a similar relative deficiency of color, as compared with the tanks that were stirred. The same holds of the single-frame fermentation, where the color is even less; and the fact that an even amount of tannin was extracted notwithstanding the pomace was in a solid mass at the top, is explained by the high temperature which, as the table shows, prevailed in that cap. The same consideration doubtless applies to the "old-style" (No. 562), in which the high temperature of the pomace-cap offset the lack of stirring, and both tannin and color were fully extracted.

A singular and unexplained fact is the deficiency of tannin in the tank with open foulage, without cover, for which no obvious cause can be assigned; the duplication of the determination, however, leaves no doubt of the fact, which can hardly be explained without assuming that some of the tannin at first extracted was subsequently destroyed by the action of the air. If this were so, the full complement of tannin in the "Morel" product might be explained by the presence of the astringent stems.

The column giving the color-intensities is very instructive also. It will be seen that those yielding a low color were the two tanks with frames, already discussed, and the low-temperature fermentation, No. 562, in which despite diligent stirring, and the pretty full extraction of tannin, that of the color remained incomplete, being nearly one-third less than the maximum.

The full discussion of the bearings of these fermentation experiments is perhaps best deferred until the development of the wines, and their full analysis in their more advanced condition, shall give more data in regard to the final results of the several treatments. Those familiar with the subject of fermentation may, however, already derive important lessons from what is recorded above. Of course, these results must be verified by repetition during the coming season, before they can be accepted as maxims; but there is much that cannot well be upset by any subsequent experiments. Among the points that may be considered well settled, is that the method of fermentation adopted by this department (viz.. floating cover, with thrice daily stirring) is amply justified by the outcome of the nine fermentations. It secures all the advantages of aeration, full extraction of tannin and color, and maximum of alcohol, without any risk of acetification if properly managed. The method has been carried out on the large scale by Mr. John Gallegos for two years past, and has yielded excellent results; the only difficulty encountered being that in the case of very soft-skinned grapes, the frequent stirring reduced them to a pulp which it was difficult to press. In such cases the stirring must be moderated and made with implements having the least crushing effect; but I am satisfied that in the hot vintage-climate of California, the leaving-open of fermenting tanks to the access of air is most objectionable, is one of the most common and prominent causes of unsoundness, and should be done away with universally, adopting either the use of floating covers, or at least a cover over the top of the tank. Whether the disadvantages of the single-frame system can be overcome by a repeated pumping over of the liquid from below over the pomace, is a question yet to be determined; but that in the use of this method there is always a serious loss of color and tannin can hardly be doubtful.

E. W. HILGARD.

CHAPTER V.

D'HEUREUSE AIR TREATMENT.

Alleged improvements, involving sweeping changes in many industries, should above all bear the light which close practical investigation may shed upon errors to which new as well as time-venerated doctrines are subject; corroborative tests only can establish their value. Theories, apparently sound, by neglect of some essential condition, may fail to be confirmed by tests. One of the most general and firmly rooted notions has been the dread of access of air during fermenting or preserving operations; the most satisfactory proofs only will establish the fact that exclusion of air should be abolished, and that the suggestion of air-treatment as a safe, cheap, easy and effectual agent for wine and other industries is well founded. For the sake of our American wine industry it is proposed to review the principles on which air-treatment is based, the manipulations, the advantages claimed, and the conditions to be observed, the knowledge of which, by exercise of common sense, would enable any unbiased individual to test and judge for himself.

FUNDAMENTAL PRINCIPLES.

One broad principle underlies nearly all applications of air-treatment for the purpose of imparting stability, and to prevent deterioration in organic substances by a rapidly oxidizing and eliminating action on the albumenous parts, which all crude organic substances contain, so that by ordinary elementary exposure the substance may decompose or decay and thus form, in the admirable economy of nature, sustenance for other organism. The presence of the albumenous parts is an essential condition of decomposition, their removal insures stability, comparatively or absolute. Currents of air passed through the substance to act uniformly on all parts, effect first of all an oxidation of the albumenous matter, which is rendered insoluble and thus eliminated either during fermentation, by which the sugar is converted into alcohol, or by absence of fermentation at a temperature above 135° F., at which organism is killed, or by both modes in conjunction. These few plain intelligible facts constitute the whole basis of air-treatment, the applications are simply deductions.

It is certainly an error that *all* albumenous matter coagulates at a certain high temperature; if this were correct, a fluid so heated for hours could, if clear and limpid, contain no albumen. Experience plainly contradicts this, for instance, in vegetable or animal extracts obtained by heat, malt or grain wort, saccharine juices, crude oils, fats, etc. Nor do these and other substances, containing gluten or albumen, acquire stability by mere heating; if, after cooling, the germs of micoderms in the air find access, they cause fermentation or decay, as long as they find albumenous parts to feed upon. These however removed, no micodermic

action can take place, and stability is imparted. It thus becomes plain that all manipulations and processes for the
preservation of organic matter should go towards freeing
them from the albumenous parts, otherwise they remain imperfect and unreliable.

The alcohol of wine is more inclined to turn into acetic
acid the less alcohol is present, and the larger the proportion
of gluten. Thoroughly fermented wine generally contains
but little gluten, so that the heating process (to 121°–131°
F. to kill the micoderms) as a rule forms a protection; however, if not previously fermented dry, the wine will remain
sweet, for no known process but fermentation alternates the
sugar.

In accordance with the foregoing, wine freed from gluten
by air-treatment should have received full protection against
future disturbance, and the results obtained corroborate the
assumption; not only after but during fermentation, a security is obtained which heretofore was wanting, this most important part of all wine making, the fermentation, placed
under the control of time.

THEORY OF FERMENTATION.

A brief allusion to the principles of fermentation may be
in place. Alcoholic fermentation is the result of the presence of certain micoderms, that require air for vigorous
healthy action and propagation, while they suffer from want
of atmospheric oxygen. Ozone is formed by rapid passage
of air through (aqueous) fluid, and invigorates the alcoholic,
acts destructively on other but injurious micoderms that
cause disease, putrefaction and acidification; but both kinds
can only vegetate where they find *gluten* to live. Periodical
currents of air through a fermenting fluid accelerate, insure
and perfect the fermentation by invigorating the alcoholic
micoderms, and the excess of gluten is at the same time
gradually removed by oxidation, so that none remains to

support the micoderms that necessarily perish when their functions—conversion of sugar into alcohol—is accomplished. Unless invigorated, fermentation proceeds very slowly toward the end, as when the atmospheric oxygen is exhausted, the alcohol formed seems to stupify the micoderms; a large proportion of alcohol, or presence of alkaloids, (hops and glycerine for instance) prevents or retards fermentation.

Ground taste in wine is due to the gluten it contains and improper treatment which brings out the taste. Air-treatment removes with the gluten all tendency to ground taste.

HOW AIR-TREATMENT IS APPLIED TO FLUID.

To impregnate a fluid with a gas, we admit the gas below, that it may rise upwards through the fluid. Accordingly the air, to act most effectually, is admitted into the fluid in a divided state by perforated pipes or mouth pieces, sunk near the bottom of the vessel, impelled by an air-force pump. Air-treatment of a hot fluid for purification from albumen requires a vigorous and continuous current of air frequently for hours to coagulate all albumenous parts ; for fermentation however, periodical gentle currents are sufficient.

TO WINE ESPECIALLY.

A vigorous fermentation has been found the most satisfactory for must as well as other mash, and a sufficiently high temperature (75° to 85° F.) is essential. Fluctuation of temperature should be avoided as always detrimental. When the must is warmed (in a gathering tank) to about 65° to 70° F., the tanks or casks filled, the temperature steadily maintained, air is impelled vigorously for some ten minutes, and unless sugar is added, a foaming up by a rising scum, will soon take place. After this subsides (from six to ten hours) air is gently impelled two or three times each day for about five minutes at a time, till the fermentation is finished, which is accomplished at the stated temperature in from five to four-

teen days without fail. The air pipes (of block tin) are introduced into the casks through the bung-hole, in tanks from
above, and, where pulp is worked for red wine, should be
stationary for the operation. The tendency to clarify appears
at once when the carbonic acid gas ceases to form. A few
days later the still somewhat turbid wine may be drawn off to
settle in casks, bunged up, and a few weeks later will be
found clear, of free ripe taste, subject to no after fermentation
or other wine disease, free of ground taste, and fit to be
shipped to any part of the world, without more risk than old
well stored wines. Scrupulous cleanliness, sweet vessels, etc.,
are, of course, always essential.

With proper care and judgment, all wines can be quickly
finished by air-treatment, that were previously but imperfectly
fermented in the usual mode, even diseased wines (if free
from acetic acidification) restored. But no general directions
can in these cases guarantee success to careless or inexperienced persons. It should, however, be born in mind, that
in all the above cases, the object is the removal of the excess
of gluten by a quickly started and lively (though brief) fermentation, for which the presence of some sound amd active
ferment, sufficient sugar, proper heat and air are essential.
Addition of ferment may, therefore, be required, or of sugar,
the determination of the proper proportions of either, is the
work of experience and judgment. The ferment—if sound
wine yeast or another kind—should be brought into full vigorous action before it is added to the quickly-warmed wine
(of 70 to 75 F.), this temperature retained unchangeable
during the few days of subsequent finishing fermentation. All
subsequent processes are, however, obviated by the use of
air-treatment of the new must, cider, etc., which is thus carried at once beyond the reach of the many vicissitudes to
which wines fermented in the usual manner, are subject.

Wines may be classified in a general way as *sweet* and *dry* wines, or those still containing sugar, and those entirely or nearly free from it. Dry wines form the bulk of the product of European and domestic vintages, as the modes to manufacture them appeared more simple. It has been shown above how dry wines are more quickly, safely and cheaply obtained by air-treatment.

SWEET WINES,

However, are as yet obtained by partial fermentation, interruption of this process ; and by addition of spirits (to 20 per cent. or more of alcohol) or glycerine, etc., stability is imparted, further fermentation and deterioration rather kept under than precluded. They are cordials rather than wines. No sweet and light alcoholic wines are in the market (except sparkling), for the simple reason that *they could not be manufactured with any degree of stability ;* the remnant of the gluten prevented it. Air-treatment furnishes an easy solution to this question also, and permits the manufacture of sweet wines of any desired alcoholic strength and most perfect stability, because free from gluten. Americans are fond of sweet wines and should have them.

The must, fresh from the grape (or other fruit), heated to above 140° F. is vigorously air-treated for a couple of hours (or less), till the albumen is coagulated, which is removed by bag filtering, still hot. After cooling to 70° F. it is subjected to air-fermentation with the addition of a quantity of green must, suitable to insure the desired proportions of alcohol and sweetness. Or any wine obtained by thorough air-fermentation may be sweetened with crushed sugar free of gluten to suit the taste, without danger of future disturbance.

BRANDY.

The described air-treatment for fermentation of must or piquette secures by full attenuation of all saccharine parts a higher yield, of 10 to 15 per cent. of spirits, than the usual

mode, in which 2 per cent. or more of the saccharine from
the 12 to 20 in pulp or juice, is left unconverted and irre-
claimably lost. (The loss in grain mash thus saved is still
higher, from 16 to 25 per cent.)

The azotized parts are rendered insoluble, and by clearing
or straining are kept out of the still, permit the formation of
none or very little fusel oil, so that a purer spirit at once results. Air-treatment in the still during distillation of any
pulp produces spirits free of fusel (at least the first run), and
subsequent air-treatment of any distilled spirits at a raised
temperature in suitable close vessels communicates quickly
the properties of age, destroys the fusel-oils. To retain the
fullest natural wine flavor in brandies, redistillation for refining should be avoided as much as possible, and air-treatment
provides the best means to effect at once cheaply what many
years of storing is generally made to accomplish with enormous expenditure.

It is obvious that must, deprived of gluten, (what no other
known process accomplishes) in the hot state as explained,
like any other extract, may be subsequently concentrated,
without the addition of sugar or anything else, kept in casks
on draught, as preserve, confectionary, or may be employed
as addition in wine making in distant parts, to produce
greater variety of wine at any place. Enormous quantities
of thus purified concentrated must from California or other
southern grapes, containing little bouquet and much sugar,
could be more profitably employed to blend with green musts
of northern strong flavored grapes, deficient in sugar; than
turning either into wine separately.   ·

CENTRALIZATION IN WINE INDUSTRY.

If we recall to mind numerous home industries only a few
generations back, for instance the flax grown on the family
field, woven on the family heirloom, and taken to market
periodically to be sold, we wonder at the slow, tedious,

penny-wise business, that aimed to do all the work, but earned little. Mills now buy the flax, and sell the linen to the dealers. All other industries were remodelled in the same manner; producers, manufacturers and dealers are distinctly separated to make them pay ; and still we see wine men adhere to the primitive policy in wine making. As long as wines had to be stored several years to be ready for shipping, the excuse was not unfounded that the investments of distinct establishments were enormous. Air treatment, however, annihilates this objection, permits the cellar to be cleared a few months after the vintage, to be ready for next season. Central wine houses in grape-growing districts are bound to be profitable, to take the place of the numerous press houses, purchase grapes by contract for years ahead, and a few months after vintage turn over their ripe, matured product to the dealers. Large establishments work cheaper, can have more intelligent and competent supervision, have a choice of numerous varieties of grapes, to blend and produce choicer wines than the small producer can, and make business easier, more agreeable and more profitable on all sides, by yielding quick returns to all parties interested. The American wine industry can only prosper, by employing quick ripening methods in manufacture, and division of labor as indicated.

SHIPPING GRAPES.

Grapes more qualified for the table are produced in large quantities and offer better remuneration to the growers to ship to distant markets than to the press house. A great deal, however, is now spoiled in transportation and storing. It behooves us ro reduce the loss to the lowest figure. Everybody has observed that confined air favors and quickens decay; that currents of air preserve. This demonstrates the benefits of air treatment without direct oxidizing action in the gluten. Through the compartments of the railroad

car, the storehouses or vessels, currents of air are directed
with occasionally the vapors of a little burnt sulphur, or
other disinfectants, are employed with air to destroy the
germs of mould or decay that may have found their way or
even have attacked to the grapes or other produce.

A blower, run by hand or power, furnishes air or other
gasses to a system of pipes at the bottom of the compartments, and the air, after it circulates over the objects in the
compartments (a number of which can be operated in turns)
is allowed to escape by flues, or by these may return to the
blower, to repeat its action.   All kinds of fruit, produce or
meal can be preserved for a long time, at any season of the
year; which permits an exchange of the products of our
country, aye, of the globe, heretofore unattainable.

CONCENTRATED PRESERVES.

The preparation of juices or extracts, purified by air treatment and concentrated, was alluded to under sweet wines.
Many thousand tons of fruit will annually find their way into
the markets in this condensed shape, and with great benefit
to all concerned, while in the distructable green state they
would have remained almost worthless.   The aromatic, fruity
flavors are mostly retained by conducting the process of purification and concentration at a heat not exceeding 140 to
150 F.   There is no necessity to put up these articles in
air-tight, hermetically sealed jars or cans; barrels answer
the purpose.   Nor need those alkaline powders and lyes,
under the name of preserving powders or fluids (every one of
them detrimental to the digestion of the consumer) be added
to the air-purified preserves ; their keeping qualities are secured by deglutination.   It will be borne in mind that the
object in employing the aforesaid injurious adulterations, is
to neutralize an acid action for some time, by which alone
fermentation or putrefaction can take place.   Glycerine, oil
or hops, or other essential oils, even sugar, salt, alcohol, etc.,

are employed for the same purpose; that is to act, for the time being in the capacity of an alkaloid, and to retard or prevent impending changes. The mere mention makes it plain, that innumerable articles could be reduced to the fluid, syrupy or solid state in the manner described, for the sake of economy, to prevent loss by spoiling in transportation or storing, to reduce the freight by decreasing the bulk, and to return the refuse to the soil as manure, after it served as food for animals. All this, and much more, will be generally adopted before many years pass by.

RETROSPECT.

The foregoing attempt to demonstrate the importance of air treatment for the American wine industry, and to foreshadow some of the changes which it is bound to effect; equal changes by the same powerful agent, the support of all organism, are certain in numerous other industries. The revolution worked by Bessemers air process in the manufacture of iron and steel, is but the forerunner in the manufacture of organic substances of almost any kind by air treatment, a revolution, however, pregnant only of unalloyed benefits to the whole human family. In this progressive spirit I hope it will be received by those millions whose health it will secure, and whose labors it shall lighten, be it in wine making, brewing, malting, distilling, sugar or oil making, tanning, or the manufacture of extracts, transportation or storing, the purification of spirits from noxious fusil oils, or of plain drinking water from organic contaminations.

To prevent misconstruction it should be stated that the inventor of air treatment is far from considering the details, as here described, rigid rules for all cases alike; but mere details, which, according to the species of must, treated in the hands of intelligent experts, will give satisfactory results. For instance, when a few weeks more time for fermentation is no object, one vigorous air treatment of the must for one half or one hour at 60 F., previous to fermentation, may answer; in other cases, one fourth to one half hour preliminary vigorous action, and subsequent gentle treatments during fermentation, several times for one or two days, or once every day, may do the work. It must necessarily be left to the discretion of those qualified to do the work, and inclined to systematical experiments, to ascertain the best modes applicable to the varying conditions that exert their influences during fermentation. It seems essential to accelerate the fermentation, so a to carry the musts as quickly as possible through the fermenting rooms, which are frequently but poorly protected against cold, generally not arranged for heating, and of limited capacity. Proper air treatment performs this; abler men may develop more.

R. D'HEUREUSE.

The last remarks were evidently written as applicable to the Eastern industry. Here we have to guard more against excessive heat during fermentation than against cold. It will be easy for the intelligent reader to form his own conclusions, and vary his practice accordingly. Aeration, no matter by what means, or how applied, underlies all sound fermentation, and while we must aerate our must during fermentation to bring about a perfect fermentation, and deposit the gluten and albumenous substances, we must also *exclude* the air, as near as possible, as soon as the wine is thoroughly fermented and finished. I cannot think of any better simile to illustrate this, than to compare the must to a living organism, constantly omitting and *exhaling* noxious substances. When it has become *wine*, it *inhales* ; and is apt to be affected by all outward influences. There is no absolute period of quietude, but a constant change for better or worse, according to the treatment it receives ; so it behooves us to see that we treat it *well*. In this climate, we need not resort to the practices of Dr. Gall, and Petiot, so necessary in Europe and even in the

East. Our genial climate will give us a must, rich enough in sugar, and light enough in acid, to make a good, saleable wine every season, and it would be foolish indeed to resort to additions of sugar when the pure grape must is so much cheaper. While I think the practice of using pure grape sugar is perfectly harmless in Europe and the East, and even necessary to make a good, sound wine, to use it here would be folly, as we can make it without such additions, and furnish a cheaper and better article thereby. Here again, California can excel the world.

DEFECTIVE FERMENTATION.

It is or rather has been, frequently the case in this State especially in some seasons, when the summers were extraordinarily dry, followed by very hot weather during the vintage, that wines, especially the red, were "stuck" as the common expression is during fermentation; that is, fermentation set in very violently, running up the temperature in the fermenting tanks to over a hundred degrees, then suddenly stopped, when the must yet retained from three to eight per cent. of free sugar. In 1885, this was especially the case, and perhaps one-sixth of all the wines in certain sections did not "go through" as the common expression is.

The cause of this can perhaps be found in the long period of drought, when the grapes at last ripened suddenly and rather unnaturally, with many shrivelled berries, especially in the Zinfandel. The product was sluggish, and the fermentation properties not sufficiently active to carry fermentation through evenly and correctly. Add to this very hot weather, and the mistaken idea which seemed to prevail, that the most *rapid* fermentation was also the most *thorough*. The mash, generally confined under the abominable perforated heads, or still worse, left exposed without stirring, rapidly rose to a degree of heat above, which amounted to more than boiling, killed the germs of fermentation, and turned the sugar into caramel, insoluble in fermentation. The lower part of the tank was far below the upper in temperature, air was entirely excluded, and the result was wine which obstinately remained sweet. I may state with some gratification and pride, that under my method of frequent stirring, aeration and keeping the temperature even from top to bottom, I did not have a single case of imperfect fermentation, and produced as good wines as any in the market, sound and well fermented.

The reader will perceive, that in this case, as in most others, an ounce of preventative is better than a pound of cure. The most simple means to prevent such occurrences are the following:

1. Do not let your grapes get over ripe, but pick them when they show from 22 to 24° on Balling's scale.

2. Have an even temperature from 65 to 75, not exceeding 80 in your fermenting room.

3. If the grapes come in too hot, let them stand over night, and crush in the morning, when cool.

4. In fermenting, stir frequently and thoroughly, so as to aerate the whole mass, and equalize the temperature.

5. Should you not be able to work your grapes quick enough, and they run over 25°, reduce to 25 by an addition of water to the mash, before fermenting.

If you observe all these simple rules, you will have no trouble in fermentation.

But if, by some oversight or other, a cask or tank should get "stuck," or refuse to go through, the simplest remedy is to take fresh grapes of certain light varieties, for instance Burger or second crop Zinfandel, crush them, and throw the refractory must over them without delay, taking care to mix it thoroughly with the fresh grapes. This will incite fresh fermentation, and if you work the whole mass thoroughly, or aerate it, they will generally "go through" without trouble.

This is better than all the remedies suggested by wine doctors: brewers' yeast, flour, tartaric acid, plaster and tannin; and will give a better and more natural wine than any of them.

But you need not fear if you observe the above rules, that such will be the case, unless you are entirely unable to regulate the temperature of your winery. If this is the case, remedy the defect in some way or other before the next vintage is upon you. In six successive seasons, in which I have made wine in this State, I have not had a single case of imperfect fermentation yet, nor need my readers have it, if they will work rationally and carefully.

Since writing the above, over two months ago, another season of difficult fermentation has passed, and millions of gallons of wine have obstinately remained sweet. While I have nothing further to add to the advice already given, the account of experiments by Prof. Hilgard, given below, will fully confirm the views given, and serve to throw much additional light on this important subject.

Complaints of difficult fermentations have been very general during the vintage just passed, and a great deal of red wine especially has refused to "go dry" within the usual or any reasonable limit .of time. It has long been my conviction that in the vast majority of cases the difficulties complained of arise from excessive heat during and particularly at the beginning of fermentation. At the end of last year's vintage, a number of comparative fermentations were made at the University Viticultural Laboratory, partly with a view to testing this question; but it being late in the season, the only grapes available for the purpose, viz: second-crop Zinfandel, were not of a character to test the point, having high acid (.65) and low sugar (21.6); and the high temperature attained seemed to accelerate, rather than retard, the fermenting process. This season, sixteen fermentation experiments, parallel with those of last year, have been made, and the re-

20

sults of some of these throw so much light upon the causes of " difficult fermentations" that it seems proper to give publicity to them in advance of any detailed report on the whole series.

Equal charges of 200 pounds each were fermented in 50-gallon tanks, save that in the hot fermentations 25 pounds more were used, in order that the rise of temperature might be favored by greater mass. In the hot chamber a temperature of between 85° and 90° was maintained; while in the fermenting-room in which the other charges were being treated, the temperature was kept as nearly as possible at 75°. The grape employed was a fine lot of Carignane, courteously donated for the purpose by A. J. Salazar, Jr., of Mission San José. The must showed 25.75 per cent. by spindle and 53 per cent, or a little over five *pro mille*, of acid.

Of the tanks in the fermenting-room filled with mash at 63°, three, treated by usual methods, went practically dry and were sent to press on the seventh day; the first to finish being the one with "floating cover and twice-daily stirring," the method adopted in the laboratory for general purposes. The highest temperature reached by any of these was 95°.

On that day (7th) the two tanks in the hot chamber, which had in setting been warmed up to 86° and at first fermented most violently, and in forty-three hours attained a maximum temperature of 106°, had come down to very slow movement; the actual solid contents were found to be a little over 12 per cent. It being obvious that they would not "go through" under existing conditions, the two charges were divided into four parts, of which one was left in the hot chamber and treated as before, in order to observe the outcome. The others served for experiments to test the best mode of reviving the fermentation in the lower temperature of the fermenting room.

One portion received $1\frac{1}{2}$ per cent. of pomace, freshly pressed from one of the other tanks, and well stirred in ; floating cover put on and well stirred three times daily. Fermentation soon revived and went on slowly, but steadily, until the seventh day, when the charge was sent to press, practically dry.

One, a double portion of 19½ gallons, was mixed with 2½ gallons of condensed Zinfandel must set at 21 per cent. with distilled water, and having been allowed to pass into active fermentation before mixing with the "stuck" mash. Fermentation soon set in and slowly but steadily carried it to dryness on the 17th day, being 8 days from the time the fresh must was added.

The fourth portion was left without any addition but was from the time it left the hot chamber vigorously aerated, by means of an air pump three times a day. Fermentation soon revived, and the charge went dry and was sent to press at the end of the 6th day, from the time it was removed from the hot chamber, being nearly two days in advance of the other tanks treated with pomace and must respectively, but aerated only by ordinary "foulage," with cross-peg stirrer.

It thus appears that simple aeration, without the addition of any new yeast, was at that stage of the mash that had "stuck" in consequence of overheating, the most effectual mode of reviving and completing the fermentation. The pressed wine had the same acid percentage as the original must, and is free from acetic taint.

As for the portion that remained in the hot chamber, it continued a feeble action for some time, but on the fourteenth day from the setting of the mash it had practically stopped. It was then removed to the fermenting-room, and after cooling down to 75° and aerating by the pump, a faint revival of fermentation took place for thirty-six hours. Then the cap sank and the tank was "dead." The day after, the odor and taste of milk-sourness became so patent that the mess was sent to press with over 9 per cent. of solids, as a dead failure, on the seventeenth day ; a woful, but in practice but too familiar example of the results of hot fermentations.

I reserve for the future a detailed discussion of the subject, in connection with other experiments, but the main points illustrated may be briefly thus stated :

1. While musts of low sugar contents and high acid may be successsully rushed through to dryness at a high temperature and make a sound wine, the same is not true of those having high sugar and low acid ; the margin of difference between the two cases is a very narrow one, both as to temperature, acid, and sugar, and hence a few days of hot "norther" may easily turn the scale.

2. When the temperature has not been excessively high and not maintained too long, simple aeration by means of a pump or blower may revive it at a lower temperature. Sound pomace, or fresh fermenting must, are additions to be used when available or necessary.

While these facts and principles are not new to experts, I have thought it worth while to re-establish them by facts and figures, and to offer them as a substitute for the supposed mysteries of "difficult fermentations" that have so vexed our winemakers. The vatting of hot and over ripe grapes and the omission of proper aeration of the mass, while allowing the surface to acetify, are responsible for nine-tenths of all unsoundness in California wines.

E. W. HILGARD.

BERKELEY, November, 17, 1887.

(d) LIQUEUR WINES OR SWEET WINES. CHAMPAGNE.

This may be said to comprise all the sweet wines, also fortified wines. These may be made naturally, by leaving the grapes on the vines until over ripe, when, if the must is over 28° Balling, it is apt to retain part of the sugar unfermented.

This is done in Europe in several ways, either by letting the grapes hang on the vines until very ripe, and the small berries are half dried, in some cases even picking out the ripe berries with needles, and then exposing them to the sun for several days, upon screens, or straw; they are then crushed and pressed. The must of course, being so very rich and syrupy, will take a long time to ferment and develop, longer, very likely, than our impatient people would be willing to wait for them. It is in this manner that the celebrated Tokay is made in Hungary.

As I have never had much to do with making sweet or fortified wines, I shall not go into any very elaborate descriptions of the process, which come hardly within the province of the smaller wine producer, for whom this book is calculated. I shall only refer to the methods in a general way, especially as I do not profess to be a judge of these wines, nor partial to them. In fact, I do not consider them *wines*, in the true sense of the word, which with* me is the pure, fermented juice of the grape. But as they are consumed to a large extent, I do not feel justified in omitting them altogether, leaving my readers to inform themselves if they wish to make them, from a more competent source than I claim to be.

Angelica or *Sweet Muscatell*. This is generally made from Muscat of Alexandria, by letting the grapes get very ripe, then crushing and pressing them, and as soon as this is done, add about a quart of grape brandy of the usual strength to each gallon of must, also stirring in about a gallon of fresh lime to each 100 gallons of the must. This suppresses fermentation, and clarifies the wine within two days. As soon as it is clear, it is drawn off into casks, which are filled: and only needs ageing to make it more palatable. The German Muscateller or Frontignau, if treated similarly, will make a much more delicately flavored wine, and it is from this grape the French make their celebrated Muscat Lunel, which sells at $3.00 per bottle. We would hardly obtain such prices here, however, even if we made it better, for it would not be *French*, nor "far fetched and dear bought!" Yet it deserves a trial, and very fine wines of a similar character have already been produced here.

Sherries and *Ports* are generally made by fortifying with alcohol up to eighteen to twenty-three per cent. Mr. Crabb adds grape syrup to his port, made by boiling down sweet must. Sherries are then kept in a heated room with a temperature of 140 to 150 F., for three to four months, a so-called oven; and thus acquire the aged taste and flavor which their admirers fancy. It would not be of any special interest to the reader to enter into a description of the Bodega and Solera system, by which sherries and ports are made and aged in Spain and Portugal, as I do not think that Californians will ever be willing to wait ten years before they can thus ripen and sell their wines, and go into the tedious process of establishing them. I believe, however, that there are many of our grapes which acquire the sherry flavor simply by aging in the cask. I have tasted Mission at Mr. Dresels twenty years old which had it in a marked degree, and which I would prefer to most of the artificially made sherries I have tried. This is especially the case with many of our white wine varieties when they get very ripe. The Sultana, for instance, develops some of it even the second and third season, and it may thus not be difficult to produce a *natural* sherry, preferable to the artificial, by simply aging the wine of such varieties, which would seem to me to be a more proper and cleaner way than exposing them to the influence of air and mold, by leaving them in casks partly full and with their bungs open, as in Europe.

Champagne or *Sparkling Wine.* There are also two methods to produce this, the so-called *natural* way, by which carbonic acid gas is developed in the bottle by adding syrup and aging it, and the artificial, by which the wine is impregnated with it in about the same manner as soda or other artificial mineral waters. Mr. Arpad Harazthy, I think, is the only one who now follows the first in this State, his "Eclipse" is well known and generally well received by the people, although many assert that the artificial is just as good and pure. I can really not see where the great difference is, as in both cases the "liquor" is added *artificially*, and both are certainly not *pure wines*, according to the true definition of the word. As long as they contain nothing deleterious to health, and the people enjoy them and are willing to consume and pay for them, they are a legitimate branch of our wine industry, and should be protected and fostered as such, whether made in theso-called *natural* way, and sold at sixteen dollars per case, or in the *artificial*, and sold at ten dollars. There are several firms engaged in making the latter to a certain extent in this State, and Mr. Werner of New York City is manufacturing it there from California wine, which is reshipped, and consumed here to some extent.

Grape Milk. The same firm has also put an article on the market which is called Grape Milk. This is simply must in which fermentation has been suppressed by a process best known to himself, and which is sold for commercial and other purposes, to those who think it inconsistent with temperance and Christianity to use the fermented juice of the grape. There are many methods of suppressing fermentation in must; sulphuring, adding salycilic acid, etc., but I do not think any of them entirely harmless, and all more or less injurious to health. I believe. if He, whose followers these men and women profess to be, saw any harm in the moderate use of pure wine, He would not have changed water into this beverage, of which the master of the feast said that the best had been kept to the last, nor would He have instituted it as one

of the Sacraments at the last supper He took with his followers. I do not think that what was pure to Him, the purest of
all, can be impure to us, who feebly try to follow in His foot-
steps. provided we use it with moderation, as it should be
taken.

CHAPTER VI.

AFTER TREATMENT OF THE YOUNG WINE. RACKING.

As soon as the must has fully gone through fermentation,
and has become perfectly quiet, we call it *wine.* In short, as
soon as fermentation has converted the sugar into alcohol, the
must has lost its sweet and pungent taste on the tongue, and
is beginning to deposit its lees, instead of throwing them to
the surface, as it does while fermenting, it may be called *wine.*
The plainest indication of this stage is, when, in holding your
ear over the bung hole of the cask or tank, you hear none of
that hissing noise which accompanies fermentation; and the
wine, by drawing a sample from the top, which can easily be
done with a small hose, or a liquor thief, does not produce
that pungent, prickly taste which characterizes it while fermenting. It is time then to close the bung, driving it in
lightly, however; in case any after fermentation should set in,
which might injure the cask. In a week or so more, it can
be driven in tight, so as to exclude the air. But before this
is done, it will be well to fill up with the must separately fermented for that purpose. Fill the cask up to the bung, either
with the wooden funnel, or a can with a long pipe, bent at the
end and made specially for that purpose. (Fig. 31½)

Fig. 31½.

If the wine is yet in the fermenting room, it can be drawn off and removed to the cellar proper, for further development; although, if well and fully fermented, it will be perfectly safe in the fermenting room. This first racking, however, had better be deferred until it has become clear, and deposited most of the lees at the bottom of the cask, which is seldom the case before two months. The old idea that the lees were the "mother of the wine," and necessary to it until February and March, is fully exploded now. The lees are nothing but the excrements of the wine, the impurities contained in the must, which fall to the bottom and are deposited there. As soon as this has been done, the wine will gain nothing by remaining on them; on the contrary, the sooner it is taken from them, the better for it.

But no matter whether on the lees or off, the casks should be kept *full;* and it is necessary that this be done once a week; from a small cask kept for the purpose. It is not necessary, however, if the casks have been filled up once, that this be done with the same variety of wine. Any good, sound white wine will do to fill up *all* the white wines; and one also for the red. Of course it would not do to fill white wines with red, as that would have a tendency to color them; but if the red wines have color sufficient, a filling up with white wines will do no harm, but rather give them finish and smoothness.

If the·wines are kept in tight casks, have sufficient alcoholic strength, and filled up every week, there is not much danger

of mould; or as some call it, very inappropriately, *flowers* on
the wine. Should it however happen, that the young wine
shows a white film on top, fill the cask so full that this film or
mould which floats on the surface, runs out at the bung, and
after it has all run out, bung tight.

RACKING.

As soon as the young wine is clear, it can, and in fact,
ought to be racked. For this purpose we need *a.* a clean
cask of about the same dimensions as the one you wish to
commence with. *b.* Faucet of sufficient dimensions. *c.* A
small tub to put under the cask and faucet. *d.* Either a
pump or buckets to transfer the wine into the empty cask.
f. If the latter, the wooden funnell referred to before. If the
former, sufficient hose to reach from faucet to pump, and
from pump to cask. Of course the empty cask must be clean,
sweet and tight, is placed where the wine is to remain until
racked again, and is laid on the supports so that the front end
with the hole for tapping is, say two inches, lower than the
end next to the wall.

Some rack through a siphon (Fig. 33) from the bung hole,

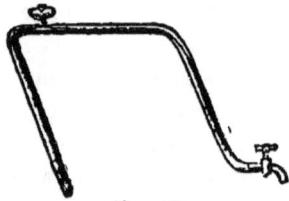

but I prefer the faucet ; as the
hole for it is just about, or ought to
be, where the lees commence, and
the bottom of the clear wine, which
cannot be so accurately guaged
with the siphon, and therefore is
more apt to disturb the sediment.

Fig. 33.

For racking large casks, it is also well to have a jack, (Fig.
34) to raise the cask when it has run down to the level of the
faucet. Now, the cask being in position, we are ready for
the operation. Loosen the bung first, by a tap or two with
the mallet, for, if this is not done, the air entering at the
top or faucet hole, will disturb the lees. Then take the

faucet in your left hand, a bucket or tub between your knees to receive the wine which may spurt out, loosen the plug with the mallet, until you can draw it with your hand, and as soon as you withdraw the plug insert the faucet, which of course should be closed, and drive it in firmly with the mallet. Now have a glass handy, and try whether the wine runs clear and limpid. If not, open the faucet only about half way, and let the wine run slowly, testing it from time to time until it comes clear. Then shut the faucet, and put the turbid wine separately into a cask or keg. It will soon clear, and can then be drawn off again.

FIG. 34.

If you rack with buckets or cans, it is a good practice for very young wine to open the faucet but partially, so that the wine comes in a spray or circle. It is thus somewhat aerated, and the oxygen will help the final clarification and tend to ripen it sooner. Avoid as much as possible to shut the faucet suddenly, the check is apt to disturb the wine, but have two buckets or cans, slipping the empty one under the faucet and removing the full one. A little practice will soon give the necessary dexterity. If the cask is high, you want an assistant to empty the cans into the funnel.

Pumps are much more convenient and not very costly. The accompaning cut shows a very convenient pump, to be had at Woodin & Little, San Francisco, and the manner in which they are operated. They save a great deal of labor, and for operations at all extensive, will pay for themselves in a very short time. (Fig. 35).

When the cask is about empty, and the stream through the faucet diminishes, try frequently whether the wine runs clear. As soon as it becomes cloudy, shut the faucet, and put the cloudy wine into the separate cask for that purpose. Should

it still be clear, when run off, shut the faucet, and tip the cask
gently, either with the jack or by hand, say six inches at the

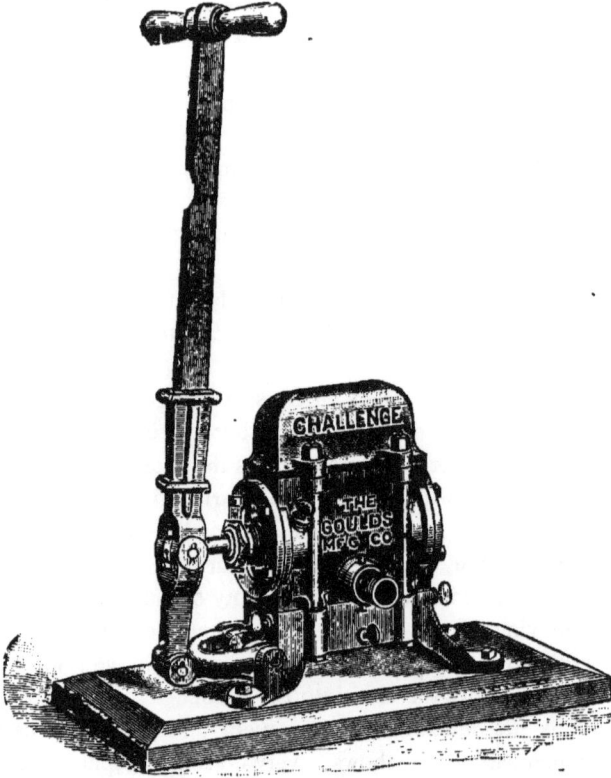

Fig. 35.

further end.   Then try again and if it still runs clear, it can
be added to the first.   The cloudy wine also should only be
used as long as it runs somewhat limpid ; as soon as it becomes thick it should be shut off.   If the wine should be
mouldy on top, it must be closely watched ; for as soon as it
runs down to the level of the faucet, the mould will run out.
This can be filtered by laying a clean flannel cloth over the

funnel, which will retain the mould. But with proper care this will only happen with very weak, light bodied wines.

Then withdraw the faucet, and let the sediment run out through the hole, or if your cask has a manhole or door, as all over 300 gallon capacity ought to have, you can unscrew it, and take the lees out. They can be used for brandy, but should for that purpose either be distilled immediately, or kept in air tight casks or tanks until all are ready.

When all the lees that will run out have been emptied, put several buckets of clear water into the empty cask, shaking it thoroughly, so that all parts are reached. Repeat this until the water runs from the bung perfectly clear and limpid. For small casks the rinsing chain is very useful (Fig. 31) as the sediment is very slimy, and if any remains on the sides of the cask, it is apt to injure the wine. Large casks with manholes, can of course be brushed clean inside. Remember that cleanliness is absolutely necessary, if you want clear, pure tasting wine. You cannot have it without this.

A great many sulphur all the casks before using them, especially white wines. It is well enough to use sulphur to keep empty casks sweet and free from mold, but they ought to be rinsed with clear water before using. Unless white wine is dull, and lacks spirit, I think sulphuring a positive injury, which is apt to destroy or at least vitiate that delicacy which should be their characteristic. If otherwise good and sound, they need no sulphur to make them so. To red wines it is a positive injury, as it deadens their color and decreases it.

Having racked and cleaned one cask, we can refill it with the next; and should the first not be quite full, fill it with the same or a similar wine, and so go on until all is finished.

The principal rules to observe are:

1st. Choose bright and clear weather, and avoid damp and rainy days, as well as storms.

2d. Do not rack until your wine is clear, unless it is to remove it to a different temperature, or you need the casks for further fermentatiou.

3d. Do not rack when the vine is in bloom, or when the fruit commences to color.

4th. Do not have the wine exposed to the air for any length of time.

5th. Take care to do it thoroughly, and keep it clear from all traces of sediment.

6th. During either very hot or very cold weather, keep the door of cellar or fermenting room closed, so as to avoid changes of temperature.

Generally speaking, racking is necessary only twice a year, if performed thoroughly and well. New wines should be racked in December, or as soon as clear, then again in February or March, and again in August. This will of course vary with the climate, and no fixed rule can be given. Unless the wine has been handled and racked very carefully, a slight fermentation takes place in June or July, and as soon as it becomes entirely quiet again, it should be racked. Very much depends on the thoroughness of the operation, when performed the first and second time.

CHAPTER VII.

CLARIFICATION, FILTERING AND FINING.

If wine is sound and well made, it seldom requires anything more but careful racking, and this is certainly preferable. If, however, from some cause it will not clear of itself, it may become necessary to do so artificially. We can do this by two different methods, mechanically by filters, or chemically and mechanically by fining.

FILTERING.

This acts simply mechanically, as the wine is pressed in some way through a substance which acts as a retainer for the impurities contained in it. Among those most commonly in use are paper filters, where a strong pressure forces the wine from above through the pores of blotting paper, also through felt and woolen bags. The one most perfect in its action, and which has at the same time the advantage of low cost, automatic action, and being cleaned easily, I have seen at the inventors, Mr. A. Beck, corner 6th and Mission sts., San Francisco. I have also seen wines which had run through it, and compared them with the same wine before filtering, and can testify to their great improvement from the process. The inventor deserves great credit for his ingenuity and skill, which has resulted in an apparatus within the reach of every one, and which ought to be in every cellar. (Fig. 37) represents the apparatus in use. The wine to be filtered is contained in cask A, which is elevated on a platform a few feet above the filter, B. The wine runs through a faucet and hose, into the bottom of the filter, which contains a number of circular flannel sacks, drawn over spiral springs to keep them sus-

pended.   The wine raises in the filter by the pressure of the
fluid from the cask above, is pressed through the bags, rises

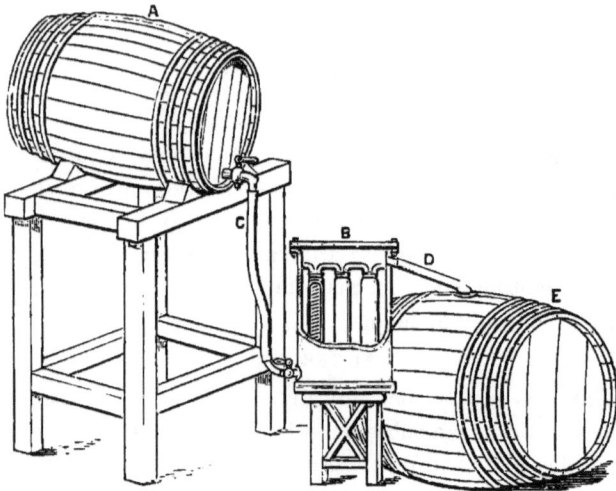

FIG. 37.

to a false bottom which holds them in position, and flows
from there through a hose D, into the cask E, below.   It
takes about 12 hours, with a filter of 10 gallon capacity, to
filter a puncheon of 160 gallons, and the apparatus, when once
started at night, needs no looking after until the next morning, when the cloudiest wine has been transformed into a
liquid as clear and bright as the sun.   The sacks can be
easily cleaned, by forcing water from above through them, or
taken out and washed for further operations.   I have been
thus explicit because I think the invention destined to super-
cede all finings as well as all other mechanical processes, and
be of real benefit to every wine maker, while its cheapness
brings it within the reach of every one.   In fact I think it so
perfect that it would be superfluous to describe other filters.

FINING WITH CHEMICAL SUBSTANCES.—FOR WHITE WINE.

Gelatine and Isinglass are the most common and best finings for white wine. The first is prepared from the bones,
skins and tendons of animals, and comes in tablets or sheets
generally. It is one of the most powerful of finings, and
takes a great deal of tannin and color with it, should therefore not be used for red wines, except when it is desired to
deprive them of an excess of tannin and color. It precipi-
tates more sediment than most other finings, is apt to leave
a bad taste in the wine, and wines fined with it should be
racked from the finings as soon as cleared. It is generally
only used to clarify common white wines, and if they are
rather flat, tannin should be used with it. Take about one
ounce for one hundred gallons, and soak a few honrs in
water. Then dissolve it in a dish over a slow fire with a
little water, which, however, should not be allowed to boil,
and stirring constantly.

Isinglass or Fish Glue is made from the bladder of the
sturgeon, and comes mostly from Russia. This is the best
fining for white wine. Take one ounce to one hundred
gallons of wine, break it up by pounding with a hammer on
a block of wood into small fragments, so that it will easily
dissolve. Put in an earthen vessel and pour enough of the
wine to be fined over it to cover it; and add a little more
after an hour or two, when the first has been absorbed.
When it has become a jelly, in about twenty-four hours, it
can be thinned by adding more wine, and working it by the
hand until entirely dissolved, then strain it through a piece of
linen, using pressure enough to squeeze out the mucilage.
It should be whipped or beaten, and more wine added if too
thick. It can be kept in bottles for some time when prepared, by adding a little brandy.

ALBUMINOUS SUBSTANCES.

Among these are the blood of animals, milk, etc., but I
21

do not recommend them, as the first is apt to leave a bad taste, and the second may cause lactic fermentation. The white of eggs is the best of albuminous substances, and is mostly used for clarifying red wines. It coagulates by the action of the alcohol and tannin, and forms a precipitate heavier than the liquid, carrying with it as it falls, the matters in suspension in the wine. Only *fresh* eggs should be used, but the yolks must be carefully kept out, as they discolor the wine. Take a dozen eggs for a hundred gallons, and beat them up by whipping thoroughly, together with a small quantity of wine before using.

For weak wines, containing so little spirit that the finings do not act, alcohol must be added. For wines that are deficient in tannin, this should be added; for upon the proper quantity of this, and the alcoholic strength, depend the action of the finings. If the wines contain enough of alcohol, as they generally do here, and the finings do not act, the cause is generally a deficiency of tannin; and sufficient must be added to produce the desired effect. One-half to one ounce to the ordinary tannin of commerce is generally sufficient. Dissolve one-half pound in a quart of strong alcohol, by shaking thoroughly in a bottle of double the size. When it has been mixed twenty-four hours it is filtered, and one gill of the solution contains one ounce of tannic acid.

After the finings have been prepared as above, two or three gallons are drawn from the cask which is to be treated, by siphon or a small hose from the bung hole, the finings poured in, and thoroughly stirred with the wine. This may be done with a stick split at the end into several prongs, or by a brush formed by bristles or flexible wire. (Fig. 38). The wine drawn out should then be filled in again, until the cask is perfectly full, and left to rest until the wine is bright. The time in which this takes place varies from two to four weeks, three weeks being about the average. But it should not be left on the finings after it has cleared, as even the best and

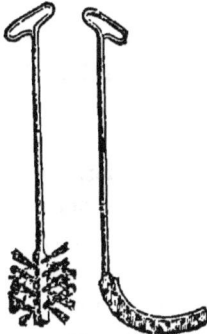

purest are apt to impart a disagreeable flavor, if the wine is left on them too long.

I trust that the filter described may do away with finings altogether, and thus save expense, unnecessary labor, and risk of any taste from the finings. We want to furnish *pure wine*, the most perfect and pure we can have, to the world; and any-thing which will enable us to do so within the shortest possible time, should be wel-comed and adopted by our wine makers as well as by the trade.

FIG. 38.

CHAPTER VIII.

AGING WINE.

We hear a great deal on the subject of aging wine, on Cali-fornia wines being too young when they are sold, etc. There is no doubt a great deal of truth in this, for it is a well recog-nized fact that strong, full bodied wines require a longer period for their full development than lighter wines; and as our wines are of the former class, it is but natural that they should require time to bring out their best qualities. But to understand this fully, we must understand first the true mean-ing of the term "old wine."

I call a wine "old" when it is fully developed, when it is perfectly clear and bright, having deposited all the impurities it contains when young, and has obtained the highest degree of perfection, has fully developed its flavor and bouquet, in short, when it has arrived at full maturity. If bottled at that period, it will retain these qualities and perhaps even improve for some time, as it is made as near air tight as possible, but we cannot expect any further improvement in casks, and it is not a remunerative article to keep after this.

Thus it often happens that a certain wine is older, that is more developed, at six months or a year, than another is or will be in three years; owing to the treatment it may have received. Let us consider the means we have to *age* a wine, in succession, and we will know better how to attain age as soon as possible.

Fermentation. Complete fermentation is certainly the first step towards complete development; without it we cannot expect to have a wine which will develop rapidly. If this progresses regularly and thoroughly, not too fast or too slow, so that the wine is dry in six to eight days from pressing, there will be little trouble afterwards. Air treatment will do a great deal to help this along, and should be applied, in red wines by frequent foulage or stirring, in white wines by keeping them in a well-regulated temperature, and if they show any sluggishness, by conducting air through them by the D,Heureuse process, racking in the manner indicated, by letting the wine run through the faucet in a spiral manner, or through the rose of a common watering pot, so that every particle comes into contact with the air. Of course, this is only to be applied while fermentation is still going on; when this is over, it would work to the detriment of the wine, instead of improving it.

Temperature. This is an important agent in aging wine, after it is thoroughly fermented. Wines kept at an even temperature, summer and winter, will improve more and faster, than those subjected to sudden changes. Hence the difficulty of properly aging wines in very hot climates.

Filtering and fining. These are important agents in aging wines, by removing all the substances which would induce a second or third fermentation.

Heating. Pasteur has invented a process by which the wine is subjected for a short time to a heat of 130 to 140 F. This is on the theory that when wine has come to a certain stage of development, the heat kills all the germs of further fermentation. This, however, excludes all further improvement also, and while it may be advisable to apply to common wines, I would certainly not apply it to fine wines.

As a general rule, if wine is well and thoroughly fermented, and well treated afterwards, in racking, clarification if necessary, the casks are kept well filled, and at an even temperature, it will not be found necessary to resort to any more artificial means to age it. Good treatment will often produce an older, *i. e.* more developed wine in one year, than the same variety, but under slovenly treatment, will furnish in three. In fact, slovenly treatment will not and ought not to make, *good* wines; while careful handling will always produce them. It is the old question of the survival and success of the fittest, which is eminently verified in wine making,

There is a great difference also in the wines as to the time when they reach their highest perfection. Wines rather thin and light, deficient in body and tannin, as also in color, will develop sooner and reach their highest state of development in a much shorter time, than those rich in sugar, flavor, color, tannin and alcohol. In fact all the grand, fine wines need a longer period to reach perfection, but will also retain it much longer than the reverse. It takes more time naturally to bring out their high quality, but when it does come, it is to stay. Thus it may be safely asserted that the general run of Zinfandels and Burgers will not improve after two or three years at the furthest, while the Cabernet Sauvignons, Chauche Noir (or true Burgundy), Riesling and Traminer will be best after two years, and keep improving as they grow older. I have tasted Rieslings and Traminers in this State, eight to ten years old, which it would be difficult to excel any where for flavor, richness and mellowness. Who ever has such wines, and can afford to keep them, will certainly not loose by doing so; while the producer of lighter wines will do well to sell as soon as he receives a fair offer. Let us not forget that it is not age *alone* which gives wines their quality and their name; there must be something else to make them grand wines. Small wines in fact have a sprightliness when young, which makes them appear better than they really are. They will sell best as long as this remains, and become flat and dull with age.

CHAPTER IX.

DISEASES OF WINE.

Here again, preventative is better than cure. Wine properly made and handled, will not become diseased in our climate, where we always have sugar in the grape to produce it of sufficient alcoholic strength to keep it. In this respect, though the French may beat us as *wine doctors*, we have the advantage of them in the perfection of our product, which needs no doctoring, if well made and treated.

But still we have patients enough in our State, made so by improper treatment, and although hardly competent to prescribe for them, as my wines were generally healthy, I will try and give some advice in cases of emergency, which may arise even in the best regulated wine cellars.

Earthy flavor. Sometimes, young wines have this, when the grapes were grown on poorly drained or very rich, or heavily manured land. The best cure for them are frequent rackings, they should not be left long on the pomace in fermenting, and as soon as passably clear and quiet, they should be either filtered or fined energetically. If rather flat and deficient in tannin, they should have some tannin added, about an ounce for 100 gallons, with the finings, which will help to deposit the insoluble matter, and then racked. Repeated rackings will do much to remove it.

Greenness. This is caused by an excess of tartaric acid, and gives a sour taste to the wine, resembling unripe grapes. It is caused by picking the grapes too soon. It need not happen here, as we can always have our grapes ripe enough, but is often found in the wines made from second crop. It will gradually disappear with age, after the first two rackings, when the tartaric acid falls to the bottom and sides of the casks, and chrystallizes there. It is sometimes the case that a heavy bodied wine, containing little acid, can be blended with such wine to mutual advantage. In this case, make a test with a small quantity first, mixing it in a glass, until you have the proportions for a blend to benefit *both*, so that you can work understandingly, taking as a rule, the benefit which the *better* wine receives from the blend, not the advantages to the inferior. Sometimes, surprising results are obtained in this way, but it takes long practice and a good tongue to produce good blends. However, each wine maker should strive to be proficient in the art, without which his practice is still incomplete. This is by far preferable to adding lime or other alkaline substances, which neutralize the acid, but are unhealthy and should not be used.

Roughness. Caused by excess of tannin, and is not always a fault, but an excess of a good quality in young wines, which will disappear in time. It can be avoided in fermentation, by taking the wine from the pomace sooner, as indicated before. If the rough wine has only astringency, without bitterness or excessive acid, it will improve very much by age, and it is safe to leave it alone. Judicious blending is often very valuable also, as for instance blending the wine which may be very rough, but contain little acid, with dark color, with one that is deficient in tannin, but has abundance of acid, and lacks color.

Sourness. Is generally caused by a too prolonged fermentation on the skins, or appears in wines that were "stuck" in fermentation. It shows acetic acid in the wine, and if this is present in any marked degree, so as to become at once perceptible to the nose or tongue, the best course is to distil such wines; they are hardly worth the trouble of doctoring, and will never be quite sound again. It is also caused by exposure to the air from looseness of the bung, and from using soured casks, which impart it at once. If all these are avoided, as they should be in a well regulated cellar, there will be no milksour, nor pricked wine. Some authors recommend neutralizing the acidity with chalk or marble dust, but my advice to the reader is, to leave these unwholesome practices alone, turn your pricked wine into vinegar and brandy, and resolve to have no more of it in future.

Weakness. We are not troubled with this here, if we plant the proper varieties. If we have some, however, the proper remedy is to blend with a heavy, full wine, or to add alcohol or grape brandy, from one to two quarts to each 100 gallons.

Flatness, Mouldiness, or Flowers. These only appear in neglected or weak wines, and will seldom be found in well regulated cellars, or in wines properly made and handled. They generally go together, and are the consequence of exposure to air. If the bung is frequently removed, and the cask not kept well filled, the vacuum becomes filled with impure air, and the wine degenerates, forming a white film or

mould on top, which the Germans call kahm, and the English writers very unappropriately, I think, the *flowers* of wine,
for it certainly does not bring forth good fruit, and is anything but ornamental. The surface of the wine which shows
them has become flat, acetic, and mould begins to form; and if
not counteracted at once the wine will spoil. It is generally
found on wines weak in alcohol, or those that are neglected
in filling up and bunging. The simplest mechanical means
of counteracting it is to fill the cask so that it runs over, and
if this is done gently and slowly the mould will float out on
the surface of the wine. When no more mould appears, the
cask is bunged up tight. It will, however, also be necessary
to rack the wine into a fresh cask which has been newly sulphured, and when the wine has run down to the faucet look
out closely; if any mould appears, keep the wine separate; or
if you fill it into the same cask spread a flannel cloth over the
funnel, and strain the mouldy wine through it. If this is done
promptly the wine can be saved, but if left in the cask with
the mould on it, it will soon spoil altogether, becomes flat,
looses all sprightliness, and acquires a disagreeable, mouldy
taste.

Dull bluish or leaden color. Flavor of the lees. This is
also due to neglect generally, and will seldom appear in well
regulated cellars. Its cause is generally improper racking, or
rather neglect of racking at the proper time, or mixing turbid
wine with the clear at racking, or irregular temperature in the
cellar, also defective fermentation. Rack into a fresh cask
well sulphured, and see if they will clarify. If they still remain dull and turbid, it is to be supposed that they lack
either tannin or alcohol. If the former, it can be added as
indicated before; if the latter, a quart of alcohol to every
twenty gallons. The alcoholometer will show if the wine is
weak in spirits, and the above proportion is for a wine of eight
per cent. of alcohol, ten being the lowest normal strength of

California wines.   If it still remains turbid, after two to three
weeks, filter or fine heavily, and when the wine is clear, put
into freshly sulphured casks.   Such wine is apt to go into
putrid decomposition, and should be watched closely ; but
well fermented and handled wines will not show any of these
symptoms, and the cellar man is generally to blame for their
appearance.

Ropiness or toughness.   Its cause is a viscuous fermentation
in wines, which makes it slimy in appearance, so that it does
not run freely, but draws in threads.   It is caused by an excess of albuminous matter, and want of tannin; generally only
appears in certain white wines, which are very mild naturally.
It can be cured by the addition of an ounce of tannin to one-
hundred gallons, dissolved in wine and added in the manner
of fining, stirring it well, and after two or three weeks, rack it.
Sometimes lack of tartaric acid is the cause, and certain of
our best wine grapes, for instance Franken Riesling, which
are rich in albumen and rather lacking in acidity.   A new
fermentation over the husks of grapes rather high in tannin
and acidity will also cure it.

Mouldy taste.   This comes from impure casks, and the
cellar man is to blame for it when it does occur.   It can only
be prevented by cleanliness.   The wine should be racked
into a sweet, clean cask, well sulphured; or fermented again
over fresh pomace; but will generally retain a trace of it.

All of these diseases seldom occur if the proper care is
taken, and I can only reiterate the instructions as to the utmost care in fermentation, cleanliness of *all* utensils, racking
at the proper time, and with proper care.   If this is done,
we need have no diseased or defective wines.

CHAPTER X.

This is an art in which especially the French excel, and which has given them such prominence in the wine market. It can not be acquired in a few days, or even a few seasons, and yet it is something that every wine maker should understand, and on which his success in a great measure depends. It depends on an intimate knowledge of each variety of the grapes he handles, its prominent qualities and its defects, and he cannot do any successful blending, before he has tried each variety separately, and knows what kind of wine it will make by itself. Nor can French and German experience avail us much here; as the varieties they use with eminent success, may give an entirely different product for us. And again, the experience and practice of Northern California cannot avail in the South, nor be alike every season, as each section and each season may and will give a different product.

There are two kinds of blending, *before* or *after* fermentation. The first is done by fermenting the grapes of two or three varieties together, picking the grapes on the same day, and mixing them in the fermenting vat, or even on the press and at crushing. This no doubt is the most natural and intimate way of making blended wine, for in fermentation the union becomes complete, and one variety often materially assists the other. For instance, we will suppose a case of two varieties, Chauche Gris and Burger. The first, when fully ripened, is very rich in sugar, very full bodied, rather deficient in acid and tannin. The Burger is light in sugar, has a superabundance of acid and tannin. The first is rather sluggish in fermentation, the last ferments easily and quickly. Here would be a case for successful blending, and there is no question that about one-third of Burger, added to two-thirds of Chauche Gris, will make a better wine than each by itself. But then a difficulty presents itself in their different times of ripening. The Chauche is medium early, and the Burger late in ripening; and even the Chauche ought not to get dead ripe, but be taken when it shows about 25° B., when the Burger ought to hang until fully ripe, at least a month later. We can, therefore, not ferment them together, but must make the wine of each separate, when the proper time comes, when each will give us a perfect product, or as near perfection as the two grapes will yield, and then mix them after fermentation. Now let us suppose another, Marsanne and Burger. Here we have a case which we can blend in the fermenting vat, because they ripen at the same time; or Marsanne and Herbemont, or Marsanne and Clairette Blanche. All ripen late, Marsanne is very full and smooth, lacking acid, with decided and very full flavor; the others are sprightly, with rather superabundance of acid and tannin. Each by itself will make a desirable wine, but fermented together with something like one-half Marsanne and one-half of Herbemont and Clairette, or two-thirds Marsanne, and one-third Burger, will in most seasons produce a nearly perfect wine; and "go through" quickly.

Another case in point. Refosco or "Crabbs Black Burgundy," as it is better known, has fine color, plenty of acid and sugar, fine bouquet, but is rather defective in tannin. Grosse Blaue, or Koelner, has little bouquet, but superabundance of tannin. They ripen at the same time, therefore can be blended in the fermenting vat. Very likely two thirds of the first, and one third of the latter would make a successful blend. But the exact proportions will have to be determined by the experimenter, as no uniform rule can be given to cover the differences which location, soil and climate may make in the composition of each.

Zinfandel will blend successfully with many varieties, as it has sprightliness, good flavor, and abundance of acid, but lacks fullness, smoothness, and in many locations, color. Lenoir, Chauche Noir and Mondeuse, also Petit Bouschet are good varieties to blend with it. A very fair guide for blending in the fermenting vat the next season, is to ferment each variety separately, and then making tests in a glass, say for instance, when you have separate samples of Marsanne and Herbemont, or Clairette Blanche, take four glasses, one with pure wine from each, and mix in the two others; take for instance one half Marsanne, one half Herbemont, and pour them together, changing them from one glass to the other, until they are thoroughly mixed. Then compare the mixture with each of the pure samples, and see whether it suits your taste better than either of them alone. If yet too full, try one-third Marsanne, two-thirds Herbemont; if too acid, two-thirds Marsanne and one-third Herbemont, always mixing well, and comparing with the pure samples. A little practice will soon enable you to find the right proportions, and when you have determined on these, you are not very apt to go far wrong in fermenting them together the next season; and the same practice will enable us to successfully blend what we already have. Sometimes, three varieties can and ought to be used to make a successful blend. For instance, for the celebrated Chateau Yquem, three varieties are used, the Semillion, Sauvignon Blanc, and Muscadelle de Bordelais. But in making these tests, they can only be of value when the different wines are in about the same stage of development.

Again, it may become desirable to blend the product of two vintages. One may be full and rich, the other light and sprightly. The same procedure will show us what to do in such cases, but remember that, when you want to blend two different vintages, each should be a finished *wine*, fully fermented and clear, as blending is apt to cause an after fermentation, if this is not the case.

The aim in all blending ought to be, *to produce a wine as perfect as we can attain*; in short, *improvement* of two really *good* wines, which, however, may yet lack certain qualities which the other does possess; *not* to make a *poor wine* barely *saleable* by, *blending* it with a *better*. If you want to attain a name and fame for your product, never sell a poor wine under your own name. If any one else can use it, dispose of it at a reduced price, or condemn it to the stile. We can always make sound, drinkable, good wine; let us resolve to produce and sell no other. Blending, if followed as indicated above, is an important factor to attain this end.

CHAPTER XI.

BOTTLING WINE.

This is rather for the dealer than the producer ; yet every producer may want to keep some of his wines in bottles, to see how they develop, and sometimes to keep small quantities when racking. I shall not go into this subject elaborately, but simply give a few brief rules, which will enable any one to keep wine in bottles, without going into the commercial part of it, which belongs to the wholesale dealer.

The wine you want to bottle should be *ripe*, that is it should be *perfectly fermented*, *clear* and *bright*, have its bouquet developed, and not leave a trace of the pungency on the tongue, which is always a sign of slow fermentation.

The bottles should be perfectly *clean*, and of *good glass*. For all wines resembling hock, take the long slender bottle, generally known as hock, for red wines and Sauternes, the common claret bottles are used. For cleaning bottles, a common brush of hog bristles, put cross ways through a handle of strong double wire, does good service, though they can generally, if new, be cleaned by rinsing in cool water.

Use good corks; for on the cork being air tight, and clean and fresh, depends the keeping of the wine in a great measure. Scald in boiling water first to make them soft, and extract all impurities ; let them stand a quarter of an hour until they are thoroughly steamed and softened, then drain off the hot water, and immerse them in cold clear water.

For small quantities, no elaborate and costly bottling machine is needed, but a single hand machine, consisting of a wooden cylinder, with a rim lined with rubber, to fit on the neck of the bottle, and lined with tin or zinc inside, will be found very convenient. The cork is put in at the top of the cylinder, which is placed on the neck of the bottle, a wooden pestle put on top, and the cork driven down into the bottle by a few blows of a wooden mallet. The bottle should be placed on a somewhat elastic substance below ; and as the cylinder is narrower below than above, it compresses the cork so that it will enter the bottle. They are made by parties in San Francisco, Mr. Henry Waas, I think ; and can also be found at Justinian Caire.

You also want a small faucet that will fit the faucet hole of your cask or barrel, and is small enough at the end to go into your bottles.

We are now about ready for the operation, provided the cask you intend to bottle from, has been placed securely, so that no stirring of any deposit can cloud the wine. Open the bung first, then place a bucket or small tub under the faucet hole, remove the plug and drive in the faucet firmly. Then test the wine with a glass, to see if it is perfectly clear and quiet. If not you will have to let it rest for a few days until it is, as it is worse than useless to bottle cloudy wine. If clear, fill your bottles to within an inch of the cork when drove in. It will expedite the operation if one will fill the bottles, and another cork them. Sometimes the wine will run cloudy at first, but be perfectly clear after a few bottles have been drawn. These should be kept separate, and will deposit their sediment in a few days, when they can be racked again. Drive the cork in to the rim of the bottle, and let the wine come to about one inch of it, after the foam has subsided. The bottles should then be laid flat on their sides, so that the wine will cover the corks. They can be laid on the floor of the cellar, or stored in bins made for the purpose, of lattice work and boards. These are generally so constructed that they will hold a double layer of bottles, which can be placed with their necks resting against each other. If securely corked with good corks, it is not necessary to seal them, though wine dealers either seal or capsule them. But as any one can easily inform himself how to fit wine for the market by labeling and packing in cases, if he intends to go into that trade, I shall not describe all the operations necessary for that purpose, but merely confine myself to the subject for domestic use, and to keep some samples of peculiar vintages, as each wine maker ought to do.

When the wine has rested for a few weeks, it ought to be examined whether it has made any deposit at the lower side of the bottle. Should this be the case, care must be taken to keep the bottle on the same side, lay it firmly on a board or table and draw the cork, letting the wine run off the sediment, which imparts a disagreeable taste to the wine. But if it was thoroughly ripe and well clarified or filtered, this will not be the case.   Red wines however, will generally deposit and loose a little of their color with age, and if well handled, can be taken off clear.

The conditions to be observed as to temperature, weather, etc., are about the same as have been described in racking, in fact bottling is racking, only into smaller receptacles, and for greater convenience in handling small quantities for consumption.

The exact *time* when wine is ripe for the bottle, cannot be determined generally, this must depend on its development. Many wines are riper when a year old and more fit for bottling, than others are at three years old.   But it is hardly safe to bottle even the most developed wine until it has passed through its first summer, as a slight fermentation is apt to set in, which must have entirely subsided, before it is fit for the bottle.   With very heavy, full bodied wines, it may take several years.   But when fully ripe, wine will develop and keep better in bottles than in wood, as there is always some evaporation through the pores of the latter, which is excluded in the bottle.

22

CHAPTER XII.

CONCENTRATED MUST.

This is one of the new industries, grown out of our necessities of finding a market for our products, and our somewhat
isolated situation, which makes economy in freight rates particularly desirable. If we can condense must to forty gallons,
where we had 150 before, the reader will see at once what an
immense gain this will be in freight. Moreover it steps in at
the right time to relieve our over production of red wines, as
mainly red wine grapes are used. If we can ship 400℔ of
condensed must to England or any part of Europe, without
danger of spoiling, instead of 150 gallons, or 1500℔ of wine,
and it can then be fermented into a good, sound claret, by
the simple addition of the same amount of water extracted
here, it will be apparent to every one that a vast amount will
be saved in freight alone ; and that we should welcome this
as one of the most timely innovations we have.

There are at present two must condensers in operation in this
State. One is the Yaryan process, of which Mr. Thomas D.
Cone is agent, and who has made a trial of the process at Mr.
Krugs cellar, near St. Helena. The plant is small, having a
capacity of ten tons a day, and costing, set up and ready for
operation, $2,500. The grapes are crushed and pressed, and
the must is conducted to the feed tank of the apparatus by an
automatic device. From here it is drawn through a lateral
coil of pipes, which may be described as a pipe within a pipe,
the grape must is in the inside pipe, and this is in a larger one
surrounded by steam. The must is then condensed by the
action of the steam, aided by a vacuum system, and drawn

into a separate chamber ; while the water taken from it is drawn into another direction in the form of vapor, and afterwards condensed and discharged through the waste pipe. The condensed must is drawn by a pump into barrels for shipment.

The grape must in its condensed form is almost as thick as jelly, and contains seventy per cent of sugar. One hundred and fifty gallons of must are condensed to forty gallons, and of course the saving of freight alone is a large item. When it reaches its destination it is fermented over again by adding the same or a larger quantity of water than has been extracted, and thus made into wine. Mr. Cone has not been able to get a full water supply at Mr. Krugs, consequently has not been able to operate to its full capacity. He hopes to demonstrate by this season's operations the entire practicability of the condenser, which would be within the means of one or several of the larger producers, who cannot avail themselves of the large condenser on the Springmuehl plan.

THE SPRINGMUEHLE CONDENSER.

A party consisting of chief ex-officer Wheeler, Mr. Charles Krug, T. D. Cone and commissioner Isaac de Turk returned from a visit to Sonoma County, where they visited the large condenser lately established one mile north of Clairville by the stockholders of the American Concentrated Must Co., J. de Barth Shorb, President. They found it in successful operation. It has a capacity of one hundred to one hundred and fifty tons per day, but as their crushing and pressing facilities are somewhat incomplete, they are only working about fifty tons per day. Only claret grapes are used, for which about twelve dollars per ton is paid. The grapes are crushed, pressed, and the dry pomace afterwards mixed with the condensed must, so that all the wine making ingredients are preserved which the grape contains, and the product is shipped direct to London, England, where it will be fermented and turned into wine. They were much pleased by what they

saw, and think that the concentration of must in large quantities will materially assist in developing the wine markets.
Dr. Springmuehl, who was there, stated that he expected next
year to put up two of these plants himself, one at Fresno and
one at Los Angeles.

The Springmuehl system is only applied and applicable to
very large quantities, as the inventor contends that only in an
apparatus of very large dimensions a perfect product can be
obtained. The Yaryan process is applicable to smaller quantities and the capabilities of individual producers. If both
are successful, they will have a very beneficial effect on our
industry, as they will open an immediate market for our surplus red wine grapes, and assist all those who have so largely
planted them without having the facilities and the knowledge
to make them into wine.

I refer here to the essay of R. D'Heureuse on air treatment
again. It is a remarkable coincidence that he should at that
time already have foreshadowed the necessities, problems, and
processes, which our decade seems just about to solve, and
his words sound like prophecies. May they be fulfilled to a
degree which surpasses his most vivid imaginings.

CHAPTER XIII.

BRANDY AND VINEGAR.

That a large quantity of brandy could and already is made here, from the pomace and lees as well as from wine itself, can be drawn from the single fact that 1,500,000 gallons of wine, of the vintage of 1885, were distilled into brandy. Not being very familiar with distilling myself, I shall not go into detailed descriptions of the apparatus and process, which had better be conducted by experts, should it become advisable to do so. The small producer had better not meddle with it, but leave it to his more wealthy neighbors, with whom he can easily make arrangements for distilling, if advisable.

Brandy can be made from the pomace and lees, but it is generally somewhat harsh and rough, and the prices for brandy have been so low of late years that it has hardly paid to utilize these. But prices for brandy have advanced lately, and it may become profitable in the near future to use them. To use the pomace for this purpose, it is generally saturated with water when freshly pressed, refermented, and the liquid thus obtained, familiarly called piquette, distilled in the usual way. To make brandy from the lees, they are thrown together into casks when racking the wine in winter, diluted with water, and also distilled. Then also, immense quantities of milksour and defective wines are distilled every year, and it is about the best that can be done with wines very rich in sacharine, but which did not "go through" in fermentation; much better than to try to doctor them up, and ruin the market by their sale as wine. It is self evident that these will make a large amount of brandy, as they contain a great deal of sugar, and the more sugar, the larger the yield of alcohol.

But the finest brandies are made from the grapes themselves
distilled in their fresh or partly fermented state, and it has
been the special study of some of our best brandy makers, to
find varieties, which would produce the mildest and finest
flavored brandy.   Mr. Geo. W. West, of Stockton, has produced a very fine brandy from a grape called Wests White
Prolific, probably a Spanish variety, but which has not as yet
been identified.   As it is also an immense bearer, and makes
a very delicate white wine, it is a desirable variety to plant,
for those who have in view the manufacture of brandy.   Gen.
Naglee, of San Jose, has made some very fine brandies, which
connoissuers contend owed their high quality partly to the
selection of varieties, partly to careful handling and ageing.

It may yet be advisable in certain of the southern districts,
where grapes ripen early, and develop a large amount of sugar
to use the product mostly for liqueur wines and the mauufac-
ture of fine brandy.   The sale of 60,000 gallons in a single
year by Mr. Rose, who has always made a very fine type of
brandy, will show that there is a large market for a really
good article.   In this direction, planters should pay more attention to the heavy bearing white varieties, such as Burger
and Folle Blanche.   The latter is used to a large extent in
France for the production of fine Cognacs.

But the smaller producers can use their pomace, or at least
a part of it, for vinegar.   This needs no costly buildings in
this State, nor expensive fixings.   A shed outside of the cellar,
but on the sunny side, and a few tanks are all that is needed.
As the pomace contains generally sugar enough, if fermented
over, all it needs is to fill them with pomace and water, and
leave them exposed to the air, when acetic fermentation will
set in in a short time, and convert the water into fine vinegar.
They should however, not put on more than about one-third
of the quantity of water, than the must which was pressed
from the grapes.   When fermentation is over, the clear liquid

may be drawn from below, or pressed. Thus, every grape grower can have his own pure wine vinegar, infinitely more wholesome than any he can buy, and as good wine vinegar is higher in price than wine, and finds a ready sale, he can dispose of the surplus at a paying figure. Imperfectly ripened grapes can also be utilized in this manner, should they not contain sugar enough for wine. In that case, no water need be added ; they are simply crushed, and exposed to the air.

CHAPTER XIV.

WINE STORAGE HOUSES.   THE "PURE WINE" BILL.

Any one at all conversant with the history and growth of the industry of this State, must acknowledge that one of our greatest drawbacks has been the immature state in which our wines were thrown on the market. There was not sufficient old, sound wine held over from the former vintages, to enable the dealers to meet the demands of the trade, and the consequence was that wines were shipped East when hardly a year old. With the large quantity of albuminous substances our heavy musts must contain, the imperfect manner in which many were made and fermented, it is not at all surprising, if, in spite of all fining and clarifying, these wines should come to their consumers cloudy and immature in many instances, thus seriously injuring the trade and the reputation of California wines. That this is a great detriment to the prosperity of our calling, will at once become apparent. Yet the greater part of our producers are not able to hold their wines until the second year. They need their

cellars and cooperage again, and must dispose of their wines
of the last vintage in some way, before the coming one is
upon them. The dealers even, often have not capacity enough
to hold large quantities, and generally buy only a few months
in advance of their shipments. Thus California wines, in
the majority of cases, come to the consumer in an immature state, without having developed all their best qualities.

The remedy for this is apparently simple and near at hand.
We have large grain warehouses in all parts of the State,
where the producer can store his grain, and obtain an advance upon it. Yet *good, sound wine*, stored in the same
manner, would offer a much safer investment to the capitalist than wheat or other grain, which is subject to the depredations of rats and mice, weevils and other insects; while
good, pure wine is not only safe from all these, but with
proper care and handling, improves and gains in value instead of deteriorating, as grain does.

It was with evident satisfaction that I visited the first of
these establishments on the Pacific Coast, the immense cellars
of the " California Winery and Security Company, " at the
corner of Brannan and Eighth streets, and saw that this long
discussed project bids fair to become a living reality. I was
conducted over the immense buildings formerly owned by the
California Sugar Refinery Company, by the obliging Secretary,
Mr. D. M. Cashin, and all the details fully explained. The
buildings have a capacity of from five to six million gallons in
puncheons, and about 700,000 gallons have been stored so
far ; of which 600,000 gallons are dry wines, 100,000 gallons
Angelica and Port. Mr. Cashin tells me that about three
millions have been engaged so far, and although this, like all
new enterprises met with many difficulties ; and the cleaning
of the building of all the old machinery, etc., occasioned a
great deal of delay, they are now fairly under way. The
building keeps a very even temperature, is four stories high,

well ventilated, and a side track from the depot of the Southern Pacific affords the greatest shipping facilities. In the equable climate of San Francisco, with a mean temperature of 60 F in the building, the conditions for ag ;ing wine could scarcely be better. The Company have secured the services of Hon. H. Pellet of St. Helena, well known as an experienced wine maker, as Superintendent of Cellars. The conditions under which they receive wine for storage are as follows :

1. The owner must send by express two sample bottles of wine to be stored, foi examination, with particulars as to quantity to be shipped, and capacity of casks required. One of the samples is submitted to Prof. Rising, State Analyst, for analysing, the other submitted to the cellar superintendent. If approved by both, the Company will supply cooperage, if required, containing 50. 60 and 160 gallons, as requested by shippers. .

2. Advances will be made on pure, sound wine only, viz.: ten cents per gallon on wine. When cooperage is required, it will be supplied by the company, on which further advances to the amount of actual cost will be provided. The cost of puncheons will be about six cents per gallon. The rate of interest to be seven per cent. annually.

3. Storage will be at the rate of twenty-five cents per ton per month, about equal to 160 to 170 gallons. Fire insurance at the rate of one per cent. per annum.

4. Racking charges will be estimated according to the labor employed, and will be about fifteen cents per puncheon per annum.

5. An estimate of all charges, interest on advances, storage insurance and racking charges, will amount to about three and a half cents per gallon per annum, and the value of the wine will doubtless increase to double its ruling rates, by its maturing under expert treatment and in an even temperature.

The owner has the privilege of fixing the price at which he is willing to sell when the wine is marketable, which will be the lowest limit at which it will be sold by the company to buyers. The advances made to him, and the storage offered, will enable him to carry on his operations, empty his cellar, and receive the benefits accruing from the increased value of the wine. On the other hand, it will enable even the dealer to replenish his stock easier, to better advantage and of more uniform quality than by the present system of making selections all over the State; and especially the Eastern and foreign trade will find it to their advantage to purchase, where they can find large and uniform quantities of well developed and matured wine. These advantages are so striking that they must be apparent to every one, and I hope that this is but the initiatory step to a general system of wine warehouses on our Coast.

In connection with this, it may not be amiss to say a few words in regard to pure wines, and the so called "Pure wine bill." I believe that the adulteration of wine has never been practiced to a very great extent on this coast, but that the bad repute in which some of our wines were held, arose more from their imperfect and faulty handling, than from real adulterations. That some unscrupulous persons used cherry juice, and even more injurious substances for coloring and smoothing over defects in some of the wines of inferior grade, cannot be doubted; but hardly to the extent which some asserted. For this, the prevailing custom of selling whole cellars of wine, good, bad and indifferent, to the merchant, and compelling him, so to say, to take a lot of trash, if he also wanted the really good wines a cellar contained, is in a great measure to blame, as much of this trash was not saleable unless doctored to some extent, and the merchant of course tried to get his money back out of it. Be that as it may, it became necessary and seemed advisable to prevent this, and make a demonstration to show the world what *we* consider *pure wine*. With this intent, and for that purpose the present law was designed, and passed our Legislature, which I insert here, together with explanations by the Chief Viticultural Officer, and opinion of Attorney-General Johnson. Its constitutionality is now being tested, and I hope it will be of great benefit when effectually carried out. Should it not be found perfect or practical, it can be amended so as to become so, and will thus help to raise the standard of our wines. The dealers will quit buying poor wines, which will then go to the distillery, and our wine makers be compelled to take more pains than has been done so far, to produce a really saleable article.

SUBSTITUTE FOR SENATE BILL, NO 219, ADOPTED IN SENATE
FEBRUARY 17, 1887—AN ACT TO PROHIBIT THE SOPHIS-
TICATION AND ADULTERATION OF WINE, AND TO
PREVENT FRAUD IN THE MANUFACTURE
AND SALE THEREOF.

The people of the State of California, represented in Senate and Assembly, do enact as follows:

SECTION 1. For the purposes of this Act, pure wine shall be defined as follows: The juice of grapes fermented, preserved or fortified for use as a beverage, or as a medicine, by methods recognized as legitimate according to the provisions of this Act; unfermented grape juice, containing no addition of distilled spirits, may be denominated according to popular custom and demand as wine only when described as " unfermented wine," and shall be deemed pure only when preserved for use as a beverage or medicine, in accordance with the provisions of this Act. Pure grape must shall be deemed to be the juice of grapes, only, in its natural condition, whether expressed or mingled with the pure skins, seeds, or stems of grapes. Pure condensed grape must shall be

deemed to be pure grape must from which water has been extracted by evaporation, for purposes of preservation or increase
of saccharine strength. Dry wine is that produced by complete fermentation of saccharine contained in must. Sweet
wine is that which contains more or less saccharine appreciable to the taste. Fortified wine is that wine to which distilled spirits have been added to increase alcoholic strength,
for purposes of preservation only, and shall be held to be
pure, when the spirits so used are the product of the grape
only. Pure champagne or sparkling wine is that which contains carbonic acid gas or effervescence produced only by natural fermentation of saccharine matter of musts, or partially
fermented wine in bottle.

SEC. 2. In the fermentation, preservation, and fortifica-
tion of pure wine, it shall be specifically understood that no
materials shall be used intended for substitutes for grapes, or
any part of grapes; no coloring matters shall be added which
are not the pure products of grapes during fermentation, or
by extraction from grapes with the aid of pure grape spirits;
no foreign fruit juices, and no spirits imported from foreign
countries, whether pure or compounded with fruit juices, or
other material not the pure product of grapes, shall be used for
any purpose; no aniline dyes, salicylic acid, glycerine, alum, or
other chemical antiseptics or ingredients recognized as deleterious to the health of consumers, or as injurious to the reputation of wine as pure, shall be permitted; and no distilled
spirits shall be added except for the sole purpose of preservation and without the intention of enabling trade to lengthen
the volume of fortified dry wine by the addition of water or
other wine, weaker in alcoholic strength.

SEC. 3. In the fermentation and preservation of pure
wine, and during the operations of fining or clarifying, removing defects, improving qualities, blending and maturing,
no methods shall be employed which essentially conflict with

the provisions of the preceding sections of this Act, and no materials shall be used for the promotion of fermentation, or the assistance of any of the operations of wine treatment which are injurious to the consumer or the reputation of wine as pure ; *provided*, that it shall be expressly understood that the practices of using pure tannin in small quantities, leaven to excite fermentation only, and not to increase the material for the production of alcohol ; water before or during, but not after fermentation, for the purpose of decreasing the saccharine strength of musts to enable perfect fermentation ; and the natural products of grapes in the pure forms as they exist in pure grape musts, skins, and seeds ; sulphur fumes, to disinfect cooperage and prevent disease in wine' ; and pure gelatinous and albuminous substances, for the sole purpose of assisting fining or clarification, shall be specifically permitted in the operations hereinbefore mentioned, in accordance with recognized legitimate custom.

SEC. 4. It shall be unlawful to sell, or expose, or offer to sell under the name of wine, or grape musts, or condensed musts, or under any names designating pure wines, or pure musts as hereinbefore classified and defined, or branded, labeled, or designated in any way as wine or musts, or by any name popularly and commercially used as a designation of wine produced from grapes, such as Claret, Burgundy, Hock, Sauterne, Port, Sherry, Madeira and Angelica, any substance or compound, except pure wine, or pure grape must, or pure grape condensed must, as defined by this Act, and produced in accordance with and subject to restrictions herein set forth; *provided* ; that this Act shall not apply to liquors imported from any foreign country, which are taxed upon entry by custom laws in accordance with a specific duty and contained in original packages or vessels and prominently branded, labeled, or marked so as to be known to all persons as foreign products, excepting, however, when such liquors shall contain adulterations of artificial coloring matters, antiseptic chemicals, or other ingredients known to be deleterious to the health of consumers; *and provided further*, that this Act shall not apply to currant wine, gooseberry wine, or wines made from other fruits than the grape, which are labeled or branded and designated and sold, or offered or exposed for sale under names including the word wine, but also expressing distinctly the fruit from which they are made, as gooseberry wine, elderberry wine, or the like. · Any violation of any of the provisions of any of the preceding sections shall be a misdemeanor.

Sec. 5. Exceptions from the provisions of this Act shall be made in the case of pure champagne, or sparkling wine, so far as to permit the use of chrystalized sugar in sweetening the same according to usual custom, but in no other respect.

Sec. 6. In all sales and contracts for sale, production, or delivery of products defined in this Act, such products, in the absence of a written agreement to the contrary, shall be presumed to be pure as herein defined, and such sale or contract shall, in the absence of such an agreement, be void, if it be established that the products so sold or contracted for were not pure as herein defined. And in such case the concealment of the true character of such products shall constitute actual fraud for which damages may be recovered, and in a judgment for damages, reasonable attorney fees to be fixed by the Court, shall be taxed as costs.

Sec. 7. The Controller of the State shall cause to have engraved plates, from which shall be printed labels which shall set forth that the wine covered by such labels is pure California wine in accordance with this Act, and leaving blanks for the name of the particular kind of wine, and the name or names of the seller of the wine and place of business. These labels shall be of two forms or shapes, one a narrow strip to cap over the corks of bottles, the other, round or square, and sufficiently large, say three inches square, to cover the bungs of packages in which wine is sold. Such labels shall be furnished upon proper application to actual residents, and to be used in this State only, and only to those who are known to be growers, manufacturers, traders, or handlers, and bottlers of California wine, and such parties will be required to file a sworn statement with said Controller, setting forth that his or their written application for such labels is and will be for his or their sole use and benefit, and that he or they will not give, sell, or loan such label to any other person or persons whomsoever. Such labels shall be paid for at the same rate and prices as shall be found to be the actual cost price to the State, and shall be supplied from time to time as needed upon the written application of such parties as are before mentioned. Such label when affixed to bottle or wine package shall be so affixed, that by drawing the cork from bottle or opening the bung of package, such label shall be destroyed by such opening ; and before affixing such labels all blanks shall be filled out by stating the variety or kind of wine that is contained in such bottle or package, and also by the name or names and post office address of such grower, manufacturer, trader, handler, or bottler of such wine.

SEC. 8. It is desired and required that all and every grower, manufacturer, trader, handler, or bottler of California wine, when selling or putting up for sale any California wine, or when shipping California wine to parties to whom sold, shall plainly stencil, brand, or have printed where it will be easily seen, first, " Pure California Wine," and secondly, his name, or the firm's name, as the case may be, both on label of bottle or package in which wine is sold and sent, or he may, in lieu thereof, if he so prefers and elects, affix the label which has been provided for in Section 7. It shall be unlawful to affix any such stamp or label as above provided to any vessel containing any substance other than pure wine, as herein defined, or to prepare or use on any vessel containing any liquid, any imitation or counterfeit of such stamp, or any paper in the similitude or resemblance thereof, or any paper of such form and appearance as to be calculated to mislead or deceive any unwary person, or cause him to suppose the contents of such vessel to be pure wine. It shall be unlawful for any person or persons, other than the ones for whom such stamps were procured, to in any way use such stamps, or to have possession of the same. A violation of any of the provisions of this section shall be a misdemeanor, and punishable by fine of not less than fifty dollars and not more than five hundred dollars, or by imprisonment in the county jail for a term of not exceeding ninety days, or by both such fine and imprisonment. All moneys collected by virtue of prosecutions had against persons violating any provisions of this or any preceeding sections shall go one-half to the informer and one-half to the District Attorney prosecuting the same.

SEC. 9. It shall be the duty of the Controller to keep an account, in a book to be kept for that purpose, of all stamps, the number, design, time when, and to whom furnished. The parties procuring the same are hereby required to return to the Controller semi-annual statements under oath, setting forth the number used, and how many remains on hand. Any violation of this section, by the person receiving such stamps, is a misdemeanor.

SEC 10. It shall be the duty of any and all persons receiving such stamps to use the same only in their business, in no manner or in nowise to allow the same to be disposed of except in the manner authorized by this Act ; to not allow the same to be used by any other person or persons. It shall be their duty to become satisfied that the wine contained in the barrels or bottles is all that said label imports as defined by this Act. That they will use the said stamps only in this State and shall not permit the same to part from their possession, except with the barrels, packages or bottles upon which they are placed as provided by this Act.   A violation of any of the provisions of this section is hereby made a felony.

SEC. 12. This Act shall take effect and be in force ninety days after its passage.

This law goes into effect and becomes operative on June 5th, 1887.

In section ten of the above law will be found the following:

" It shall be their (those employing the stamp) duty to become satisfied that the wine contained in the barrels or bottles is all that said label imports."

As there are many dealers who will employ the stamp on wines, bottled or packed by them in small packages ; which wine they receive from others in larger packages, coming to them covered by the State stamp of purity, the question arises as to the liability of such bottler and what would constitute in the eye of the law, the "duty" of the said bottler in determining that the wine employed was true to label.

The answer to this question has been kindly furnished to me as follows, by Attorney-General Johnson :

SACRAMENTO, May 19th, 1887.
J. H. WHEELER, ESQ.,
204 Montgomery St., San Francisco.

Dear Sir:—Answering your inquiries as to the Act to prohibit the sophistication and adulteration of wine, &c., approved March 7th, 1887.

You make a hypothetical case for my opinion: " A buys an adulterated wine from B, with a pure wine stamp over the bung. A bottles the wine and puts the pure wine stamp on the bottle, believing the wine to be pure. Subsequently the wine is found to be not pure. Is A then liable ? "

It won't do for A to trust implicitly B or his stamps. The Act requires some diligence on A's part. It says that it shall

23

be his duty to become satisfied that the wine contained in the barrels or bottles, is all that said label imports as defined by this Act.

A therefore must not be guilty of criminal negligence. That would be as bad as if A's intent was to palm off adulterated or impure wine.

But if A makes a reasonable effort in good faith to satisfy himself that the wine is all that the label imports and is satisfied after using due diligence, he would not be guilty of a misdemeanor, if he was mistaken or imposed upon. It is the good faith of A and the use of due diligence and scrutiny in his investigation, which the law requires. I do not think an analytic test is necessarily required to be applied by A. That might not at all times be practicable. But he must recollect that there is a duty cast upon him to satisfy himself by available and reasonably reliable means that the wine is what the label imports, and he must be satisfied.

An analysis, however, would be the most satisfactory way to test the wine.

Very Truly Yours,

G. A. JOHNSON, Attorney General.

Other than this the law seems to be sufficiently clear to need no further explanation. Particular attention is called to Section 6, which renders the sale of anything purporting to be wine — in the absence of a written agreement to the contrary — void and the vender liable for damages if it be not pure as specified in the law. According to the framers of this law, this, whether it bears the pure wine stamp or not, is the effective clause, and coupling with it the liberal recompense to the informer and the prosecuting attorney, we have incentive sufficient to greatly facilitate its enforcement.

WINE ANALYSIS.

Whenever it becomes necessary or desirable that a wine be analyzed for the benefit of a dealer, vine grower, or any person whatsoever, pursuant to the enforcement of the above law, a sample of the same may be sent to the Secretary of the Viticultural Commission, by whom an analysis will be procured from the State Analyst and a ready report made as to its purity. The machinery for this latter work was obtained in an Act passed by the State Legislature entitled:

An Act to Provide for Analyzing Minerals, Mineral Waters and other Liquids, and the Medicinal Plants of the State of California, and Foods and Drugs, to Prevent Adulteration of the same. Approved March 9th, 1885.

This law provides that the Governor of the State shall appoint one of the Professors of the University of California, as State Analyst, whose duty it shall be to analyze all articles of food, drugs, medicines, medicinal plants, &c., manufactured, sold, or used in this State, when the same shall be properly submitted to him. The law then prescribes the methods by which the samples of various articles shall be obtained and submitted for analysis, and specifies that the Board of State Viticultural Commissioners shall have the privilege of submitting to the State Analyst samples of wines, grape spirits or liquids or compounds in imitation thereof for analysis, as follows:

Any person desiring an analysis of such products may submit the same to the Secretary of the State Viticultural Commissioners, who will transmit them to the State Analyst in the manner prescribed. The analysis shall be made and the certificate of the same shall be forwarded to the Secretary of the Viticultural Commission. This certificate, as the law reads, shall be held in all courts of this State, as prima facie evidence of the properties of the articles analyzed by him.

Thus it may be seen that there lies within the reach of every wine maker or dealer, an easy means of obtaining without expense, uncontrovertible evidence wherever fraud is supposed.

Unfortunately, this law appropriated no money for the car-

rying on of the work required.   Realizing the importance of
such a bureau and its maintenance, however, the Viticultural Commission has shared its endowment with the State
Analyst and will continue to do so in order to lend all of the
aid they can to the support of the law.   Pursuant to the requirements of the Act, W. B. Rising, Professor of Chemistry
at the State University, was duly appointed State Analyst.
An assistant has been employed, and he is now ready for and
engaged in the examination of wines, the purity of which
can be quickly determined and the report made available in
a few days after delivery of the samples to our Secretary.

Concerning the expense of maintaining the State Analyst's
Bureau, it is hoped that the Board of Regents of the University,
in their manifest desire to aid the cause of viticulture, and in
view of the liberal endowment made them in the last Legislature, will come to the assistance of the Commission in the
support of the analytic work.

THE STAMPS.

Section seven provides that on application the necessary
stamps shall be furnished by the controller.

Here again, the Legislature made no appropriation with
which to purchase the plates necessary for printing the stamps
and the liberality of the Viticultural Commission is drawn
upon, they having consented to supply the first cost.

One hundred thousand stamps have already been printed
by the State Controller, and will be ready for distribution
when needed.   Their cost will be $1.50 per M. with ex-
pense of delivery added.   The stamp for bottles may be
easily affixed thereto, that placed over the bung of a barrel
will need the protection of a piece of tin such as is ordinarily affixed to the bung of a barrel previous to shipping.

The following committee of vine growers to see to the enforcement of the law has been appointed by H. W. McIntyre,
President of the State Vine Growers' and Wine Makers'

Association: Hon. M. M. Estee, Napa; J. B. J. Portal, San Jose; Capt. Chamon de St. Hubert, Fresno; J. H. Drummond, Glen Ellen; H. A. Pellet, St. Helena; Jacob Schramm, Calistoga; H. A. Meriam, Los Gatos; B. H. Upham, San Francisco; A. Erz, Anaheim; Julius P. Smith, Livermore. To these others will be added soon. This committee will proceed to collect miscellaneous samples of wine found throughout the city and state, which, if proving spurious, will be turned over to the district attorney and the case submitted to the courts.

The pure wine Act, at the time of its passage by the last Legislature was the subject of considerable criticism and dispute.

It was discussed at great length before the public, but, having ended in adoption and approval, it is to be hoped that the little inconvenience it may make a few will be amply compensated for by its good effect on the general industry. If it opens the way to any fraud we may be sure this clause will be used by the enemy; to counteract the effect of which every good feature of the law must be brought into requisition.

Many demanded the use of certain materials in preparing wine, which materials to them seemed harmless, but which had to be denied in order to exclude other more damaging adulterants. For example, we may well afford for the market within our own State to abandon the use of ordinary grain spirits in fortifying wines if by the law we are able to stop the extensive and unhealthful practice of stretching produced by the same means.

Nor must we forget the effect of this Act in enlarging the demand for grape spirits, and thereby causing the distilling of poor wines, which would otherwise be fortified by neutral spirits, and usurp the place of better wines. By the Act a native spirit is substituted for an imported one.

The healthful effect of our wines and a consequent increased local consumption, will be greatly promoted by substitution of grape spirits for cheaper poisonous spirits.

Several have already indicated their intention of using the State stamp on small packages. On bottles, particularly, will the practice be adopted, and here it will be of use. When the public demand the pure wine stamp on the bottles, as they will do when knowing its value and finding some merchants who employ it; the trade will be forced to supply them pure wine, and that under a California label. A large portion of the native wine sold in bottles, goes today to the public with a ficticious label of foreign import. The presence of the stamp will bring to public recognition the name and trade mark of California producers, where heretofore the bottle has been branded " Chateaux La Rose," " Chateau Margaux " or " St. Julien."

Whether the use of the stamp on large packages going out of the State will be harmful or otherwise, the use of the same on bottled wines must certainly result in good.

It has been suggested that the Controller's list of those dealers making application for the stamp will prove a valuable directory for the use of purchasers. It is to be hoped it may.

Dealers have now had ample time for working off their suspicious products, and every opportunity has been given those, who—perhaps with honest motives at first—have been forced into the use of cheapening processes by harmful competition, to start anew on a fair, square basis with an easy redress from others competing by dishonest means.

Whatever effect this law may have, it will aid in forming a standard of excellence founded on quality and not alone on price. The latter has proved the ruinous measure by which our wines have been gauged and marketed, a measure by which they have been caused to degenerate rather than advance. Good wine continues to improve and will pay to keep. Poor wine will rapidly deteriorate, and if not allowed to be drugged, must go either to the distillery or be made into vinegar.

If this law becomes effective, which depends mainly on the patronage of the wine-drinking public, it will double the California market for good wine. Its success will send the poor wine to the distillery, the sale of which has dragged down the price of the better product to ruinous figures; figures which preclude the possibility of marketing any choicer grades for the general public.

J. H. WHEELER.

Chief Viticultural Officer.

CHAPTER XV.

WINE STATISTICS.

These are very difficult to obtain in this busy State, where everybody seems to have his hands full, and seems to be unwilling or unable to attend to anything else. But a few items which will give a general idea of the magnitude of the industry may be of interest to my readers, as they will show the rapid increase from small beginnings.

The Secretary of the State Board of equalization, Hon. E. W. Maslin, reports the entire number of acres in vines in the State, by counties from the report of the County Assessors to be 121,440 acres, distributed as follows :

Alameda, 3,451 acres ; Amador, 846 acres ; Butte, 247 acres ; Calaveras, 1,440 acres ; Colusa, 506 acres ; Contra Costa, 3,000 acres ; Del Norte, 4 acres ; El Dorado, 1,570 acres ; Fresno, 10,185 acres ; Inyo, 95 acres ; Kern, 45 acres ;

Lake, 985 acres; Los Angeles, 17,000 acres; Marin, 493 acres; Mariposa, 500 acres; Mendocino, 108 acres; Monterey, 500 acres; Napa, 14,431 acres; Nevada, 235 acres; Placer, 2,221 acres; Sacramento, 6,465 acres; San Benito, 110 acres; San Bernardino, 9,165 acres; San Joaquin, 1,739 acres; San Luis Obispo, 275 acres; San Mateo, 625 acres; Santa Barbara, 527 acres; Santa Clara, 9,423 acres; Shasta, 147 acres; Siskiyou, 4 acres; Sonoma, 21,638 acres; Stanislaus, 498 acres; Sutter, 430 acres; Tehama, 4,972 acres; Trinity, 20 acres; Tulare, 1,229 acres; Tuolumne, 890 acres; Ventura, 800 acres; Yolo, 3,191 acres; Yuba, 165 acres.

The Secretary of the State Viticultural Commission, Mr. Clarence J. Wetmore, however, thinks this estimate altogether too low, and estimates the number of acres, from information received of the vineyard owners direct, at about 150,000. The assessors have neglected in many cases to give the number of acres for table, and market, and for wine. As far as reported from about forty counties, there are 13,760 acres of table grapes, and 59,036 acres of wine grapes.

In this connection, the wine product of the State for the past ten years will be of interest. It is as follows:

In 1876, 3,750,000 gallons; 1877, 4,000,000 gallons; 1878, 5,000,000 gallons; 1879, 5,000,000 gallons; 1880, 8,500,000 gallons; 1881, 7,000,000 gallons; 1882, 10,000,-000 gallons; 1883, 8,500,000 gallons; 1884, 15,000,000 gallons; 1885, 9,000,000 gallons; 1886, 18,000,000 gallons.

The crop of 1887 is estimated at about 16,000,000 gallons; although there is a largely increased acreage, the crop was cut short in many sections by frost and coulure, and still more so by the prevailing hot weather and drying winds during the vintage, which caused the grapes to dry up and yield much less juice to the ton than in preceding vintages; from one hundred and ten to one hundred and twenty gallons to the ton being the average, against one hundred and forty to one hun-

dred and fifty gallons last year.   Of this crop all will not be
merchantable wine, on account of defective fermentation, and
about 12,000,000 gallons may be taken as a fair estimate of
sound wines ; the balance will have to be made into port and
sweet wines, or into brandy.   Of course, it is too early yet to
make a correct estimate, but this may be taken as the ap-
proximate result.

Besides, the decrease will be mostly in the counties which
produce the finest light table wines, Sonoma, Napa and
Solano.   Napa, which produced something like 4,000.000
gallons last year, will not produce much more than 2,000,000
this year, so that those who produce choice wines, will find
a ready sale for it, and, from present appearances, at remunerative prices.   The reports during the last three years,
in the six months ending June 30th, will also throw some
light upon the increasing consumption of California wines.
They are by sea and rail, as follows :   1885, 2,181,996 gallons ; 1886, 3,227,354 gallons; 1887, 3,624,390 gallons,
showing an increase of about 1,500,000 gallons since 1885.
As it stands now, according to the nearest estimates that can
be made at random, the home consumption is about 5,000,-
000 gallons; export trade, 4,500,000 gallons; for brandy,
1,500,000 gallons ; total, 11,000,000 gallons, which would
not leave much in first hands, perhaps not more than is not
fit at present to ship, but should have more age to make it
really saleable.

CHAPTER XVI.
WINE AS AN ARTICLE OF COMMERCE.

I was very reluctant to say anything about this subject, and hoped to obtain an article from a gentleman in the trade, who is more versed than I can be. As he is prevented however, from contributing, I am compelled to do the best I can from what information I have been able to gather from the trade. I shall quote from such sources available to me, and take pleasure in presenting an extract from a circular of Messrs. J. Gundlach & Co., one of the oldest and fairest firms in the trade, regarding last season's vintage. They say, Nov. 1886, "one of the most successful vintages recorded in the annals of the California wine industry has just been terminated and we take great pleasure in submitting to our friends our views of the result, and a condensed report of the present and prospective condition of our wine market.

"The weather, during the entire season, proved as favorable as could be wished for. No early or late frosts ; no damaging winds, coulure, grasshoppers or other unforeseen mishaps retarded the development of the grapes, and our vintners enjoyed all the advantages of picking, crushing and fermenting under the most beneficial atmospheric conditions. Our "musts" indicated from 23 to 26 per cent of sugar, with well-proportioned amounts of acid, and give promise, therefore, of speedy development into elegant wines.

"The Burgundies, Zinfandels and other Clarets, show fine color, (being in some localities probably not as intensely dark as last year), but they are faultless in fermentation and in every other respect. This observation can be made in all wine districts from North to South. California's cellars never

represented a finer selection and better fermented assortment
of young wines, than at the close of the present season, and
the future will undoubtedly mention this bountiful vintage as
the *famous* year of 1886.

"In purchasing grapes, wine-makers started reluctantly at fair
prices, but competition soon compelled prices as high as last
year. Choice varieties were readily contracted, and generally
sold at very satisfactory figures. Ordinary grades were left
to take care of themselves and our Brandy Distillers had a
splendid opportunity to replenish their deficient stock, and
considerable brandy will enter our Bonded Warehouse during
the next few months.

"We feel at liberty to make the following estimate of this
year's vintage :

Napa County	4,800,000	gallons.
Los Angeles and San Bernardino Counties	4,200,000	"
Sonoma County	3,100,000	"
Fresno and San Joaquin Counties	2,000,000	"
Santa Clara and Santa Cruz Counties	1,700,000	"
Contra Costa and Alameda Counties	1,200,000	"
Sacramento, Tehama and Solano Counties	2,000,000	"
Placer, Yuba, Yolo and El Dorado Counties	500,000	"
Total	19,500,000	gallons.

"A portion (about one-seventh) of these 20 million gallons
(in round numbers) has already been or is rapidly being transformed into brandy. The production of sweet wines has
been considerably restricted ; prevailing prices appear to offer
very little inducement for this branch of our industry. Ports,
sherries, etc., will therefore not be very plentiful. The proportion of red and white wines will probably be as two to one.

"The abundant crop of light wines of 1884 has gradually
found its way into the hands of the trade, and they seem to
be well appreciated. At the present time, absolutely no
stocks of any consequence of '84's and '85's remain in grow-

ers' cellars. San Francisco merchants and shippers control the bulk of old stocks, and prices rule steady. The wines of 1885, rich and full in body and color, are developing slowly, and will be late, therefore, in entering the general market for consumption.

"The year's business has been very satisfactory, showing an increase of about two million gallons over last year's export trade, and indicating even better progress in our local California trade and coast shipments. *

"Great fear and apprehension have heretofore been entertained of over-production. The steady increase of our vineyards, productiveness of soil and climate threatened to overbalance the healthy equilibrium of supply and demand. But, in spite of prohibition and fanatical temperance agitation in some of our States and the reluctancy of Congress to protect *pure wine* against imitations and adulterations, we are making progress in every direction — we carry no surplus of accumulated stocks—our vineyardists are as active and stirring as ever, and we all are confident of continued success. Lower prices, cheaper rates of freight, a very noticeable change for the better in the average quality of our wines, and above all, their indisputable purity ; all these facts will act as powerful agents towards a rapid extension of our market and the general distribution and introduction of California wines."

This, together with the statistics given before, will serve to show the importance of wine as an article of commerce. But in addition, it is widening its sphere of consumption every year, new markets are constantly opened, and the old ones increase, just as a stone, dropped into the water, increases the circle of its commotion, so does *good* wine extend its market everywhere, where once introduced.

Our shipment lists now show as markets for our wine, all Eastern cities, Central America, Mexico, Panama, South America, Germany, Japan, Honolulu, Tahiti, Belgium and

England. Many of these are but small at present, but they help to swell the aggregate, and if the wine proves satisfactory they will rapidly increase; California itself consumes about 6,000,000 gallons annually, and if we take into account its rapidly growing population, it is safe to predict that the home consumption will reach 10,000,000 annually, within five years from now. The increased area in vines has been nearly altogether of the choicer kinds; consequently, with increased knowledge in making and handling it, we can expect a much better average product than we have had so far. Even the low prices ruling this year may prove a benefit in the end, as it has forced many to keep and store their wine, thus giving the superior product of 1886 time to ripen and improve by age. Ours is not a product that will deteriorate and loose, like wheat and other grain, hops, or even fruit. It is well known and proven that long sea voyages improve wine and develop it, consequently there is less risk than with almost every other article of trade.

France, whose vineyards were her main stay, now does not produce enough for home consumption, and must rely on Italy, Hungary and Spain, for the supplies to keep up her export trade, and—on the wine doctors art. We are now steadily encroaching on her trade on this continent, and there is no reason why we should not be able to supply all the foreign importations. Thousands of gallons of California wine are now sold annually under foreign labels. May we not confidently hope that this will cease sometime, and that California's wines will be sought, just *because* they are Californian! There seems good reason for the fulfilment of this.

CHAPTER XVII.

WINE AS A TEMPERANCE AGENT.

This may seem a strange heading to our total abstinence people, who see in wine only an enemy in disguise, not quite so intoxicating but therefor all the more dangerous, than whiskey or brandy. Yet the greater part of their aversions would vanish, if they would for a moment enquire into the condition, morally and physically, of the nations in Europe who use wine as their daily drink, and those who use distilled liquors largely. They would find sobriety, health, good temper and merriment prevailing in those countries where wine is the daily drink ; and desolation, physical ruin and wretchedness among the lower classes, where distilled liquor is about the only consolation of the poor ; a wretched one indeed, I grant them, but still too often resorted to, to deaden the feeling of utter despair and desolation of the outcasts of Society.

Then their objections arise in some degree from misconception. They have learned to know as "wine" only those deleterious compounds, which are brought on the market as Sweet Catawba, Angelica, Port and Sherry, often only mixtures of logwood, syrup, and alcohol, and which are about as strong and far more injurious than whiskey or brandy; sweetened so as to disguise the bad liquor they contain, and which bring an overdose of dullness, headache, and intoxication to all the unfortunates who may partake of them. Let me be understood, once for all, that I do not call such stuff *wine*, and that it is as different in its effects from pure light wine, with only ten per cent. of alcohol, its pleasant acidity, fine flavor, and enlivening and invigorating effect on body and mind, than quinine is from a fine Mocha or Imperial. I am referring, when I speak of *wine*, to the "cup that cheers, but not inebriates," provided always that it is used in moderation, as sensible men and women should use it. I speak of the wine our Savior Himself consecrated and ordained to be used at His holy supper; and we may point to Him with just pride as the most illustrious wine maker, to promote innocent hilarty at the wedding of Cana. Our temperance friends certainly forget this, when they claim to be His followers, yet condemn in the same breath the cup which He has anointed to be used to perpetuate His memory on earth.

But it is not to Him alone that we can refer. St. Paul says to Thimothy, "Use a little wine for thy stomach's sake, and for thine oft infirmity." And I could refer them to many more passages in holy writ to show how fallacious their theory that the scriptures forbid its use. Luther, the very type of vigorous and independent manhood, *the* great reformer of the church, says, "Who loves not wine, women and song, remains a fool his whole life long." This offers some excuse for those who forget all justice and right when they try to deprive *all* of its *use* because *a few abuse* it. Surely, Luther did not go down to a drunkard's grave, nor was he ever accused of drunkeness; and I know of not one of the eminent reformers who interdicted the use of wine, as these would-be "latter day Saints" do.

"To the pure all things are pure," and he who uses wine in moderation, good, pure wine, such as I try to show my readers how to make and handle, will certainly have too great a respect for one of the choicest gifts of Heaven to mankind to abuse it. We have brought up a family of six, four girls and two boys, all grown now; they have had free access to wine from babies up, and drink it daily ; but none of them have ever shown a disposition to abuse it, although the oldest· son has had charge of a cellar for over four years, and is now cellar master over the white wine department of Gov. Stan-

ford's vineyard at Vina. But I know many hopeful young
men, whose parents anxiously guarded them against taking a
glass of wine, and forbid its use, who now, when they are no
longer controlled by them, are becoming drunkards and sots
on bad whiskey and other abominations. It is the old, old
story of the "forbidden fruit being the sweetest."

None of our so called temperance, (rather total abstinence,
however, which is far from temperance) can be more averse
to drunkenness than I am. I hate drunkenness, and I de-
spise and pity its victims more than words can tell. But because I would like to see this great nation a *temperate* one also, I firmly believe that the total abstinence fanatics are
wrong, and that only by the moderate use of pure light wine
we can ever hope to make it so. Human nature is so constituted that it craves a stimulant of some kind to keep up its
faculties in the wear and tear of daily business life. This nation has more dyspeptics among its business men than any
other, England, perhaps, excepted. And why? Because
our business men use no moderation; they work until nature
is exhausted, then rush off to lunch or supper, swallow a glass
of whiskey or brandy hastily at the counter of some bar, and
then rush back to their business. Is it not a natural consequence of such a life that they feel the need of stimulants,
use them at first, find temporary relief and excitement from
them, and end by repeating their doses too often, by ruining
their stomachs, and go to the grave before their time, or
rather at a time, when the business men of other nations begin to enjoy life.

If, instead of this, we had restaurants which were wine
houses at the same time, where our business men could obtain their meals, take their friends with them, have a choice
bottle of wine with their lunch, and take it with and at their
meals, when it would help by its pleasant acid and other enlivening qualities to digest it, take at least half an hour for

social chat, and throw away business cares, it would be a rest and recreation to them, instead of a mere process to keep body and soul together, as it is now. I am glad to see that some of our restaurants and hotels have already commenced this; and that a fair bottle or half bottle of wine can be had with their meals. But we also want houses which keep *the best* of California's products, at a good price if you please; so that every true Californian can invite a friend from abroad, whom he may have with him, to a glass of wine which is a true representative of our industry, and to which he can point with pride as the product of our State; a State of which we have such just reason to be proud, as the noblest and the brightest, *the* State where milk and honey flow, and where every one can sit " under his own vine and fig tree."

There are two crying evils in this State yet, however, which ought to be abolished as soon as possible; and such wine houses as before mentioned would do a great deal, and prove the initiatory steps to their eradication. One is the senseless and altogether unjustifiable practice in our saloon, to keep about the poorest and lowest priced wine they can obtain, a wine that costs them not more than 35 cents per gallon, and sell it to their customers at 10 cents per glass, taking good care to furnish the smallest glasses they can get besides, so that what costs them 35 cents, is retailed to their customers at $6.00 per gallon. These exorbitant rates deter nearly every one from drinking wine, a bottle is broken into, stands for a day opened, and if the wine had any quality before, it is apt to loose it entirely, before another is willing to pay ten cents for such stuff. This brings our wine into bad repute, and works directly against its consumption and use.

The other is the prevailing custom of " treating." Four or five men are called to the bar by a friend to " take a drink," for which he pays. The others feel under a sort of moral obligation to do likewise; and so five or six drinks are swal-

24

lowed, very often against their inclination, but simply because
no one wishes to appear mean or stingy, and some of them
feel "elevated" as they call it, before the round is fairly
made. Many a young man is thus led into the paths of dis-
sipation, because he does not "want to be behind." If we
had good, respectable wine houses or wine rooms, where they
could sit at a table, have a bottle of wine between them, or
several, of which each paid his share, they would derive more
enjoyment, it would cost much less, and they would depart
sober. This is the custom in Europe, and the people are
better contented and more sober over it. Treating is an
American custom, but one of which we have little reason to be
proud, and which, as a progressive people, we ought to
abolish.

The saloons which act directly against our interests as before mentioned, we should leave severely alone. We can
never hope to be a prosperous and sober people, as long as
we let them take the advantage of us, and of the wine drinking community, in such a manner. In Napa valley, nearly
every one can buy from the producer good, wholesome white
wine or claret, in five gallon kegs, at twenty-five cents per gallon, or five cents per bottle. Let each family take a five-gal-
lon keg, bottle it, and drink it at home, instead of paying $6
per gallon to the saloon keeper for wine, or for stale beer at
$3 per gallon. It is as cheap as tea or coffee, even cheaper,
they can all enjoy it at lunch or supper, and it will certainly
do them more good, and be better than the average they get
at the saloons. And when it is so abundant and cheap, so
easily attainable, there is no cause for its improper use or
abuse, the children will learn to drink it as part of their daily
diet, and there will be no craving for more than is good for
them. Wine, unlike other alcoholic liquors, does not, in its
natural light state, create the craving for more, as whiskey or
brandy does. It quenches thirst, helps digestion, enlivens

and invigorates. A glass of wine early in the morning, I have found an unfailing preventative against Malaria, have used it for thirty-five years now, in preference to quinine or whiskey, and with better results.

I have worked in the cause of true temperance, as I understand it, which is "moderation in everything," nearly a life time, and if I did not believe that wine was the best temperance agent that we can employ, but apt to lead to drunkenness, I would quit the business at once, so great is my desire to see the American people the greatest and the freest on earth, which they can only be if they become also the most sober nation. I see in the free and moderate use of light wine the only help and salvation for us, and the best element to further us in every progressive step we take. I have grown up among its people, helped to defend it in times of war, and hope to die on, and be buried in, its free soil. To contribute my mite to its true progress, has been the dream of my youth, the fond ambition of riper years. And now, when the snows of lifes winter begin to gather on my head, I still look forth to our beloved industry to help make it what it seems to me destined to be, the soberest, the freest, the happiest, and the greatest nation on earth. It may be but the dream of an enthusiast, and I may not see its fullfilment, but I have abiding faith in the ultimate result. Let me not be misunderstood, however. I am with the promoters of *true* temperance in every *just* and *legal* means they can apply, to prevent and abolish drunkenness. I am in favor of the most stringent laws against it. Let us make drunkenness a *crime*, and punish it as such. Let every man who is found in a state of intoxication, be punished as a criminal against the laws of God and mankind, let the finger of scorn and contumely be pointed at him as an outcast from the [society of decent men and women, and the vice made a by-word and reproach throughout the land, as a disgrace on the fair face of humanity.

But I deny that you have a right to punish the innocent with
the guilty ; a right to control the free and sober citizens of
this republic from the enjoyment of a pure healthy glass of
wine, beneficial to them, and necessary to their well being,
because some of their neighbors get drunk on bad whisky,
and even on wine. Where would be our boasted liberties, if
one man was punished for the crime of another? Can I be
held responsible for my neighbor's actions, and should I be
punished because he makes a beast of himself? Out on such
narrow mindedness, which can bear neither the test of common justice and fairness, nor of our constitution. Every prohibitory measure is an infringement of justice as well, as of
the supreme law of the land.

I wish to see every man or woman a free agent, subject
only, if he or she infringes the laws, which ought to be and
are, based on fairness and right, to be punished according to
these laws. And let me tell those, who wish to shackle us by
binding them according to their narrow prejudices, that if
they desire really to see this a sober and prosperous nation,
they can sooner gain their object by making this nation a
wine drinking community, then by all the cast iron measures
they may invent to compel an abstinence, which will only
lead to greater excess. Teach the nation the proper *use* of
light wines, instead of raving against their *abuse*, and you will
do more for the cause of *true* temperance, in five years, than
you can do by prohibitory laws in a century.

CHAPTER XVIII.

THE FUTURE OF THE INDUSTRY.

I do not claim to be a prophet, nor the son of one, therefor all predictions of the future must be guess work to a certain extent. But when I look at the incipient beginnings scarcely forty-five years ago, with no knowledge of varieties, mode of treatment and culture, no knowledge of wine making here ; and then consider the results already obtained during that short period, I cannot but feel the brighest hopes for the future. We have the finest climate in the world for the growth of grapes, we are sure of their maturing every season ; we know of no total failures, we are gaining in experience every year, our skill and knowledge increases, we have American ingenuity and enterprise, with the industry of the German, French and Portuguese ; in short, the intellects and muscular strength of *all* nations engaged in the busines ; — why should we not make it a grand success, and claim the world for a market, as the French have done for their clarets and champagne, the Germans for their Hock and Moselle, the Spaniards and the Portuguese for their Ports and Sherries ? We can make all these, and as good as they can, if we only apply ourselves to the task. We have an immense territory adapted to the successful growth of the vine ; our raisin makers have proven already that they can produce as good raisins as any country, and we can raise better, more attractive and cheaper shipping grapes, which will keep better with proper treatment than those of any country, and for which we have a market on this continent. What should hinder us from becoming the greatest grape growing nation on earth ? France had reached a production of one billion five hundred million gal-

lons per annum, until the phylloxera devastated her vineyards
for all of which she found a market, at home and abroad.
What is our annual production say 15,000,000 gallons, compared with this? Yet we claim that we have more territory
adapted to the successful growth of grapes than France. Yet
we talk of over production, and we may have it until necessity compels us to put forth our best energies and intellect, to
develop new markets and extend the old, to produce *the best
of every class*, until we can proudly say, that we are ahead of
them all, as we should be. We have capital enough laying
idle in the State ; it can surely find a better and safer investment in our products, than in wheat or grain. But we must
convince them of this first. We are just now in a transitory
state ; from the first crude beginnings we emerged at once into a condition of great prosperity, when the demand exceeded
the supply, and our merchants had to buy anything they could
find, to meet it, and pay comparatively high prices for wines and
grapes. This encouraged every body, Tom, Dick and Harry,
to go into vineyards ; the capitalist thought it a fine investment for his ready cash, the laborer for his muscle, and the
small means a few years of toil had enabled him to lay back;
few of them with any practical knowledge, or an idea how
long it would take to realize. That this should create a momentary production of inferior grades, is but natural. That
it should also shake the confidence of a great many, who
looked only at the immediate results of their own thoughtless-
ness, could also be expected. Viticulture is not a mere side
show, or an occupation to fill an idle hour ; nor will success
come at the mere expense of money without thought. It is
an occupation which demands, as much or more so as any
other, close application, constant attention and work of brain
as well as hand. Thousands may and will get tired of it, drop
by the way side, and make room in the ranks for those better
fitted than they were ; but we are better off without them.

Those who persevere, who have for their motto "Excelsior," whom no momentary reverse can repress, no obstacle discourage, who love the occupation, not alone for the pecuniary gain it may bring, but who delight in the very labor it costs them, will be sure to meet with success in the end.

We have a number of new enterprises to aid our industry, which have just sprung into life. Wine storage houses, concentration of must, and a number of new firms going into the trade, with better and cheaper facilities for shipping. Some of these may not be successful at first; there are difficulties to overcome in every new industry, but they will be successful in the end, and if some should fail, because not fitted for the task, others will take it up and succeed. Better methods to make and age our wines will be found, the wants of peculiar localities and the adaptation of their products to certain classes of wine will be clearly defined, and with each succeeding year we will be better prepared and qualified for the next.

Therefore I can see no ground for discouragement, no reason why we should not say to those who are willing to make this State their home, and labor with us with hand and brain and steadfast purpose, "Come and join us;" we have millions of acres yet laying idle, which will yield their smiling returns to you, and offer a healthy and pleasant field for you and your children.

Our State authorities and State institutions see the importance of the industry and are willing to help it along. We have the State Board of Viticulture, the Agricultural College at the State University, the State Wine Growers' Association, and the numerous local clubs, all doing good work in their proper sphere, and co-operating to make our fund of common knowledge greater and more available to every individual by their instructive publications. And we have the Press with us; we must not forget, but gratefully remember that they have not failed to speak a good word for our occupation, and distributed all the knowledge they could collect, to our readers.

Why, with all these advantages, should we not reach, at the beginning of the new century, a production of seventy-five million gallons, instead of sixteen million now, and be able to find a ready market for them, while our raisins and table grapes have driven those of other nations from our markets on this continent.   They belong to us by right, as soon as we can furnish as good a product, and we ought to occupy them.   That we will do so eventually, is my firm belief; I hope that the sun of 1900 may rise on the most prosperous wineland the world ever saw, on the most prosperous, happy and sober commonwealth on the shores of the Pacific, the Golden State of California, richer in her golden wine and fruits than its mines ever made it.

Reader, my task is ended.   If this little volume, which has cost the author many an anxious hour and thought by day and night, should help to bring about this glorious result, and you should think as kindly of him, as he does of all his viticultural brethren, he is nobly paid, though he may then rest in California's soil, removed from all earthly labors.   But as long as life is spared him, it will be devoted to our noble industry, with a love that never falters or fails.

CHAPTER XIX.

WINE SONGS.

" Wine makes glad the heart of man." It is therefore not surprising that all wine producing nations have had poets who have glorified their favorite beverage in song, and that singing and merriment prevails during the vintage. But none has more of them than Germany, the fatherland of song; while America is singularly deficient in them, and as far as I know, California, the coming "Vineland" of the old legend, has not yet produced any. Let us hope that the innocent hilarity, the poetic sentiment which good wine inspires, will induce some of our poets to immortalize themselves and glorify their State and its noblest product in song. My poetic vein, If I ever had one, has long ceased to flow, or I would try. But at the conclusion of a book devoted to grape culture and wine making, I cannot forego the pleasure of quoting a few translations of German wine songs, nearly as good as the original, and hope that in the next "improved and enlarged edition," if this little volume should meet with favor enough to need one, I can have the pleasure to add some *California* lyric, by native poets.

FATHER NOAH, THE FIRST WINE GROWER.

When Noah left his floating frame,
Our Lord to Father Noah came;
He prized his pious offering,
And spake: "Thou'st done a goodly thing,
And, to reward thy piety,
Thou may'st e'en choose a boon from me."

Then to the Lord old Noah said:
" The water now tastes rather bad,
The whilst there have been drowned therein
All beasts and mankind in their sin ;
'Tis, therefore, Lord, I even think,
I should prefer some other drink."

Whereat the Lord to Eden went,
And brought him thence the grape vine's plant
And gave him counsel and advice
To tend this shrub of Paradise,
And bid him nurse it carefully;—
It pleased old Noah wondrously !

He made a solemn household call,
And summoned wife and child and all,
And planted vines, where'er they'd grow ;
Forsooth, old Noah was not slow,—
He pressed the grape and built a cave,
And put it into casks to save.

Old Noah, grateful for the boon,
Cask upon cask did open soon,
And with sincerest piety
Did empty them most willingly,
And drank yet, since the flood was o'er,   •
Three hundred years, and fifty more.

This to each prudent man does show
From drinking wine no harm can flow,
And Christian folks it warns more o'er,
No water in their wine to pour,
The whilst there have been drowned therein
All beasts and mankind in their sin.

     From the German of Kopish, translated by I. A. Schmidt.

[German Text by Gruner.   Translated by J. A. Smith.]

NOAH'S LEGACY.

When Noah felt approach his end
He said: "I'll make my testament."
He counted over all his stocks,
His cattle, donkeys, goats and bucks;
The sheep, camels, and all the rest
With which so richly he was blessed.

This done he said, " I wish to see
At once my friend the Notary."
To him, he spoke, " You shall divide
My property.   Now do it right ;
Let all my children have their share,
And take yourself what's just and fair."

Thus they divided all.   But still,
Before the lawyer signed the will,
(He was, as clerks in average,
Fond of a pleasant beverage.)
He said: "But now, beloved sir,
Who of your *Wine* shall be the heir ?"

Said Noah : " In daylight and here
We can't decide that question, dear !
Let to the cellar us descend,
And see, how there the case may stand.
Don't fear pains !"   " What my duty is,"
The lawyer said, " I never miss."

A generous man old Noah was,
And freely filled the lawyers glass.
They drew a sample every where;
They tasted here, they tasted there,
And when they had the stock gone through,
Took an inventory anew.

Back came to Noah youth and life,
He thought no more of child and wife.
" Dear friend," said he, " now put that down,

And head it with a golden crown ;
Of all the wine which here you see,
The *Human Race* the heir shall be."

No death bell! Let the goblets ring!
And jolly boys my requiem sing.
Each cask filled with the golden wine,
Shall be a monument of mine.
Write this and make, dear notary,
Eternal thus my memory."

The following acquires a peculiar significance to me, as I
look back through the past, and think of the genial spirit now
laid at rest, old " Father Muench " as he was familiarly called
by his friends. One of the pioneers of German descent, in
Missouri, who served his adopted State in her legislative halls
as well, and by his numerous writings in various fields of
literature, with all the enthusiasm of a polished and patriotic
soul, he was one of the first who followed the then new industry of grape culture, and his earliest beginnings date back
to 1846. His " American Vintners School," a text book for
the beginner, attained a deserved popularity, and was translated from the German, in which language it was originally
written. Warm personal friends as we were, I often had the
pleasure of meeting him among the vines, and at his own
pleasant homestead. At my farewell visit in 1881, he expressed the wish to " die in harness," without any previous
illness. This wish was gratified by an all wise Providence.
He was found among his beloved vines, one pleasant winter's
morning, dead, with the pruning shears yet in his hand, in
his 84th year. Peace be to his memory. One of the best
and most genial of men, he yet lives in the grateful remem-
brance of his many friends.

AMERICAN VINTNERS SONG.

BY FREDERICK MUENCH (FAR WEST.)

[Translated from the German by Mrs. Wistar.]

Plant the vine, plant the vine!
Gen'rous font of ruby wine;
In the sunlight gladly playing
Richly all your toil repaying,
Will the smiling clusters shine.

Eve and dawn, eve and dawn,
Still must find us working on,
Digging, pruning, cutting, binding,
Round their props the tendrils winding
Sweet the mete of labor done.

Sun and air, sun and air!
Leafy green, and odors fair;
Then the berries, luscious treasure
Fill the inmost soul with pleasure,
Leaves and fruit and blossoms fair.

Then at last, then at last!
Left below, our labors past,
Let us, o'er the mountains straying
Where the air's mild breath is playing
Down the vale our glances cast.

Gather in, gather in!
Let our harvest now begin;
Now the purple juice, dark, glowing,
Full and free in streams is flowing,
Young and old, come gather in.

Hear it foam, hear it foam!
Surging in its narrow home,
Let it seethe and bubble rightly
Till it sparkles clear and brightly
Here, within its narrow home.

Now come on, now come on !
For our hardest task is done,
Now we pour the wines rich treasure,
Gods might envy us the pleasure,
Clink your glasses every one.

Freedoms land, freedoms land !
'Where anew my home I planned
Lo! I drink to thee, brave nation,
Comrades, join in this ovation
Hail! our chosen fatherland.